冯·格康－玛格建筑事务所作品集

建筑设计　1997－1999

迈因哈德·冯·格康，玛格与合伙人事务所（gmp事务所）在德国建筑界起着举足轻重的作用，他们的创作无论对当代建筑还是未来即将来临的新世纪而言都是一个有力的表达。本书旨在从各个不同的审美角度来审视这个事务所近年来的作品。在创作中有时形式和功能会产生与时尚相抵触的成分，而他们的作品却再一次展示了具有社会效应的艺术魅力。

《我们工作和居住的壳》－埃蒂斯·卡塔芳格

　　本书的内容包括gmp事务所1997-1999年的作品，其中包括开姆尼兹大学的礼堂，汉诺威2000年世界博览会的基督教展馆，柏林市巴林广场的德勒斯登银行的新建筑物，此外他们还为座落在北京的德语学校进行了设计，作品中还充分体现了他们对于建设新柏林－勃兰登堡国际机场进行的设计，以及他们为汉堡—汉芬市所作出的独特规划。本书还刊登了这个创作集体独具匠心的奇妙之作—基尔·霍恩乔治的折叠式桥，这一力作是与乔治·斯克朗共同合作的结晶。本书其后登载的他们为德国新型大都会特快列车所做的内部设计，这一设计的确可谓是一种古典与现代，简约与和谐，优雅与实用的结合。这就是其魅力所在。

冯·格康-玛格建筑事务所作品集

建筑设计　1997-1999

[德] 迈因哈德·冯·格康　编著
张应鹏　王天翼　译

中国建筑工业出版社

著作权合同登记图字:01-2003-7683 号

图书在版编目(CIP)数据

建筑设计 1997—1999/(德)格康编著;张应鹏,王天翼译.—北京:中国建筑工业出版社,2004
(冯·格康-玛格建筑事务所作品集)
ISBN 7-112-06975-0

Ⅰ.建... Ⅱ.①格... ②张... ③王... Ⅲ.建筑设计-作品集-德国-现代
Ⅳ.TU206

中国版本图书馆 CIP 数据核字(2004)第 112321 号

Architektur/Architecture 1966-2000

© 2001 by Architekten von Gerkan, Marg und Partner

Chinese simplified translation copyright: © 2004 by China Architecture & Building Press, through Vantage Copyright Agency, Nanning, Guanxi, China All rights reserved.

本书经德国 gmp 公司授权我社在全世界范围内出版、发行中文版

责任编辑:丁洪良 王雁宾
责任设计:郑秋菊
责任校对:刘 梅 赵明霞

冯·格康-玛格建筑事务所作品集
建筑设计 1997—1999
[德] 迈因哈德·冯·格康 编著
　　张应鹏 王天翼 译
*
中国建筑工业出版社出版、发行(北京西郊百万庄)
新 华 书 店 经 销
北京嘉泰利德公司制版
北京顺诚彩色印刷有限公司印刷
*
开本:889×1194 毫米 1/40 印张:8 4/5 字数:210 千字
2005 年 1 月第一版 2005 年 1 月第一次印刷
定价:**59.00** 元
ISBN 7-112-06975-0
　TU·6216(12929)
版权所有 翻印必究
如有印装质量问题,可寄本社退换
(邮政编码 100037)
本社网址:http://www.china-abp.com.cn
网上书店:http://www.china-building.com.cn

目　录

8　编者的话

教学与研究建筑

19　开姆尼茨技术大学阶梯教室
38　奥尔登堡大学阶梯教室
50　莱比锡大学，人文学院
51　高级技工学校食堂 累根斯堡
53　东柏林—阿德勒斯霍夫物理学院
58　地方体操馆 圣阿弗拉梅森
59　体操馆 柏林
60　教学楼二期工程 柏林—马扎赫
62　德语学校及职工住宅 北京
71　演讲台

文化建筑

76　汉诺威 2000 年世界博览会展示厅
89　"韦瑟班霍夫二世"艺术馆
90　康斯坦丁 布宜诺斯艾利斯博物馆
94　开姆尼兹工业博物馆
95　基督教社区中心 约翰萨林 汉堡
96　纪念碑 达喀尔 塞内加尔
100　音乐厅 格拉茨 奥地利
101　社区中心 藻尔高
102　杜鹃花厅 不来梅

行政与办公建筑

106　德勒斯登银行 巴林广场 柏林
124　邦仁再保险公司办公大楼 慕尼黑
126　电信中心 苏尔
133　德国邮政通信总部 波恩

135 电信中心 霍尔扎赫斯大街 柏林
146 伯特隆技术中心 埃因根

展览馆与消费建筑

152 菲利普博览会展馆
162 里明尼展览中心
164 杜塞尔多夫展览馆
166 "新墙"商业中心 43 层 汉堡
169 波茨坦广场 柏林

交通设施

174 柏林—施潘道 火车站
187 火车站，什未林
190 勒尔特火车站 柏林
195 S-巴赫车站 汉诺威
196 泰尼勒夫机场
200 慕尼黑空港 2 号航站楼
206 斯图加特空港 3 号航站楼
209 勃兰登堡国际机场 柏林
216 折叠式桥 基尔—霍恩
220 哈韦尔河跨河铁路桥 柏林—施潘道
223 沃布利茨跨河大桥 波茨坦
224 快速列车的内部设计
228 "大都会"快速列车室内布置

政府和社区建筑

241 议会大厦 河内 越南
246 法院 安特卫普 比利时
248 市政厅 布郎什
250 警察总署 丹斯特 杜伊斯堡
253 "坏小子"史特本游乐场

258 健康中心 慕尼黑
259 爱心医院 德累斯顿大学
260 联邦政府办公楼 石勒苏益格—荷尔斯泰因 柏林
265 联邦政府办公楼 勃兰登堡和梅克伦堡—福布莫恩 柏林
268 阿拉伯联合王国酋长国大使馆住宅 柏林

城市规划

272 维也纳城市规划新结构
276 商品交易大楼 巴塞尔
277 焦油厂 不来梅
278 卡伦博格步行街 汉诺威
289 柏林—腓德烈斯汉住宅小区
292 "都市住宅" 柏林
294 阿尔斯特弗里特的开发 汉堡
296 建筑学院 杜塞尔多夫
297 贝内迪克斯广场商住楼 埃尔富特
298 2000年世界博览会 汉诺威
307 港口城市 汉堡
310 隔声墙 累根斯堡
314 住宅 舍内贝克

修复和室内设计

320 塞勒大剧院的重修 汉堡
324 哈帕格·罗伊德公司总部 改建与修缮 巴林顿 汉堡
330 戏剧院休息室大厅和衣帽间的修缮

附录

编者的话

迈因哈德·冯·格康

gmp事务所建筑作品Ⅰ 1966—1978
gmp事务所建筑作品Ⅱ 1978—1983
gmp事务所建筑作品Ⅲ 1983—1988
gmp事务所建筑作品Ⅳ 1988—1991
gmp事务所建筑作品Ⅴ 1991—1995
gmp事务所建筑作品Ⅵ 1995—1997

本期gmp事务所建筑作品Ⅶ(1997—1999)是这一系列的一个延续。从1965年创建以来，我们业已实现的设计作品中有不少迄今仍在不断地被引证。本书展示了1997—1999年间几乎所有已经实现的项目。那些暂以效果图展示的在建和处于规划阶段的项目，我们在以后完成实例后仍将补充介绍。本书将介绍20个已经实现的项目和48个仍需补充介绍的项目。

出版这个专刊系列主要有以下几个目的：
- 为我们所设计的作品作一个记录：我们是如何设计的，设计的结果与构想之间能达到多大程度的吻合。
- 维护我们所有的作品在构思及设计内容的延续性。
- 为顾客、咨询者、雇员作一个定位。
- 在建筑实体与理论要求之间作一个审核。

我们的理论定位可以参照过去几集专刊的前言，它们迄今仍是很有力度的。

在描述我们的设计方法时强调的是"对话式"的，问题与解答是互动的过程。每一项特定的工程都可能有多种解决方案，但不同的解决方案对工程的分析和估价又会对其产生一定的影响，改变它的意义和相对需求。

因此我们的设计理念不是抽象的思维平台，而是确定的行为哲学。

所以不断地论证其可信度与同时能够使250多个雇员恪守创造性实践的原则一样是完全必要的。这一点可以通过收集一定的设计和建筑作品来说明，而不是通过抽象的理论。在这一方面，专刊主要是为我们事务所提供一个信息站。在以前的专集中，我们总是要求知名建筑评论家和史学家对我们的作品和专集进行评论；而这一次有所不同，我们决定选择独立的评论家和在建筑杂志及日报上发表的文章。并且我们将在它们中筛选出与本书上所出现的建筑与项目有关的文章。在此我们衷心地感谢那些作者和出版商对此改动的支持。

博览会展馆

(BAUWELT 第3期)

当克罗普科中心和大厅8/9被登为封面后(1997/33期p.1791)，汉堡的gmp事务所赢得了汉诺威2000年博览会三期工程及天主教堂德国展馆。gmp事务所在激烈的竞争中脱颖而出，击败了汉诺威的克雷恩、里普肯·泰克；科隆的彼得·库卡以及来自柯尼斯温特的普兰纳斯格鲁特·斯特道夫、郝恩斯基、普里契、特勒；满顿的3L-设计小组。

设计提出了博览会广场

围合空间的设计构思(这是根据gmp的一个典型设计理念，1996第21期)。两个展馆的外形长75m，高16m；4m深的柱廊提供了一个从外部进入建筑的衔接部分。设计师们在展厅的北面设计了一个多功能厅，内部设有休息室和书店；在南面设计了一个庄严的大厅(长21m，宽21m，高18m)。

为了在这次世博会上打造一个"建筑精品"，事务所对材料进行了限制：表面柔和的混凝土和镀锌钢。中心大厅设计为双层立面，外层采用玻璃材料，内层采用切割得非常薄的大理石和雪花石，顶部的照明设计也更加强化了大厅的庄严感。

除了教堂的地下室(位于大厅的下面)，其他所有的建筑物都是按标准的结构系统(3.60m柱网)规划的。所以展览会后可被拆除并重建使用。事务所为这个大厅设计了一个社区中心的特色。

德勒斯登银行巴林广场 柏林

为了在柏林最具魅力的巴林广场重建德勒斯登银行的"前卫客厅"，玛格(Volkwin Marg)将进行一次过分的热情奔放，虽然有点过分，但确实是一种现代与古典结合的设计。一旦你看到其内部的设计你会感到是如此的出人意料但又是如此的赏心悦目(《建筑评论》，1999年第1期)。

如果你从蒂尔加藤到柏林市中心，就可以看到巴林广场紧邻着勃兰登堡城门，这个新的古典式的城门是18世纪西方主要传统的城门之一。广场位于城市庆典的中心轴区，欢庆游行的队伍都要从这里通过。

战前巴林广场在柏林是最辉煌的，周围有美国大使馆和法国大使馆，最豪华的宾馆(埃德隆)、阿卡德米·昆斯特宾馆以及豪华住宅区和办公楼。战后城墙、广场变戈了废墟，这里成了死亡区。德国重新统一之后，每个人都希望柏林广场重新恢复其城市广场的功能，大使馆重新搬迁过来，宾馆和艺术学院也将重新修复，同时还鼓励那些久负盛名的公司入住广场。但是对于新的建筑究竟应该建造成什么格局，各方意见大相径庭。在严格的重建要求下(P.30)，建筑物的屋檐高度必须达到22m，而且必须符合适当的布局要求。尽可能多采用石灰覆面，而且这些限制的说明也使人眼花缭乱，埃德隆就是一个未经冲淡的复制品P.25)。J.P.克莱赫兹创作了唯理论性的建筑，与战前的那些旧的作品相对而说却又异曲同工，结果落入俗套。莫尔·鲁勃将巨大的美国大使馆建在广场的西南角，使得南边的艺术学院波马等建造)黯然失色。由于对此建筑物的自由、抽象玻璃立面有诸多的反对意见，因此工程进展缓慢。德勒斯登银行柏林总部是由格泰设计室迈因哈德·冯·格康和他的合作者设计的，在本赫尼奇工地对面，占据了广场中部的位置，是迄今为止全部完工的建筑物之一。向建筑物的中部看去，它轻柔的灰色大理石建筑高高耸立，两面对称给人一种肃然起敬的感觉。一对对垂直的窗户排列在位于中间的双倍的进口处左右，给人一种

传统风格的美感。可调节的铜制遮阳板随着太阳的起落撒下不同的阴影，同时也减轻了厚厚的石墙的沉重感。屋顶仿同埃德隆的屋顶，发出耀眼的青铜色，它是设计者最推崇的部分。一旦通过铜框材料的门，一种完全无法预料的空间就突现出来，一个圆形的门厅一直通到圆形的拱顶上。上面的楼层多为办公室，而且是那些德国商人最为钟爱的小套型，人们可通过开启的窗户(这些窗户是强制设计的)随意向外张望。

这一地区的建筑必须有这样一些设施，如：隔离墙及防火墙。如果内部的办公室需要更大，那么大厦顶部的天光则可以满足日间照明。甚至一些小的部位都独具匠心，如暗灰色的钢板内层也被分割，而且节点处理精巧。为了使个别房间和中央空间充满活力，颜色也是经过精心挑选的。最值得一提的是它的螺旋楼梯，装有半透明玻璃的内部窗户和整洁的走廊，而且对于隔声设备的处理以及走廊末端自动开启的灯也都处理得很好。

谈到外部设计则更是精心选材，严格设计，使之典雅、庄严具有银行气派。周围的环境加上大楼外部坚固的岩石使人产生一种庄严重之念。当然这里不是带小孩子来玩的地方，也不是那些仅为开支票、透支的人们光顾的小银行，这里属于那种默默地追逐和操纵大宗分票的投资商们经常出入的地方，还有，巴林广场也是最适合他们居住的地方。

安尼·维娜

屋檐的延伸－哥本哈根哥 桑德布朗纳的购物中心

《柏林建筑》1998年鉴

从原则上讲，为了一个购物中心值得讨论吗？托马斯·莫尔在他的《乌托邦》一书中已经下了结论："如果了解其中的一个，那么你就了解了全部"。设计需要的是一个具有魔力的"磁场"，以及一个"停泊"之处：超市、自助店及休闲场所，这些店铺主要经营电子产品，如：计算机、录像机等相关产品，当然这些场所必须明亮。在这里还要建设一些有传统色彩的已经不复存在的商店，如：修理贵重雨伞之类的小店，像柯尼斯堡的熟食店之类。现今，这些城市角落已经从喧闹繁忙的商业区中消失得无影无踪。要想

恢复这样一些传统手工业需要冒很大的风险，但设计的主要目的在于唤回那些分散在各处的知名零售商。争论自然不可避免，焦点在于那些自给自足的小店是否能生存下去并不能主宰城市的沉浮。因此，不同的观点应声而起。

米歇尔·莫尼戈 柏林

克佩尼克购物中心

gmp事务所最具资格获得设计克佩尼克购物中心这一城市标志的殊荣。当这里的建筑物倒塌之后，面临最严峻的问题就是如何尽快地建立起一个独特的体系矗立在本是钢筋、玻璃、铝合金的原址上——一个改造一新的原德意志联邦共和国的购物中心。它的周围是艾奇、林顿公园等举不胜举的著名景点，它们分布在钢筋混凝土的丛林中，但周围这些公园都是向顾客免费开放的。商业中心的管理层声称：改进消费场所质量有利于商业竞争，商业中心的外表要和提出的口号相一致。过去那种乏味的装饰和一些从威廉港及其他的地方移来过的盆景树，其实是在用朦胧的灯光代替着日光。而现在的克佩尼克购物中心却是大不相同的，由于其巨大的体量，因此建筑具有较强的标识性。与此同时郊区的

地铁通过市区,可以在一转眼的功夫贯穿东西南北。像一个街区,又像一个古老的手工业和现代奇妙电子商品中心的综合体。你甚至可以将从设计到建造完成的过程看成一个集成电路的隐喻。从色彩上看像一朵红的发紫的玫瑰,设计者对颜色的选择来自"火山熔岩"。银色的铝合金用于装饰窗户及边缘,在gmp典型的缩进部分采用这种边框处理,使整个建筑的体量在视觉上得到了控制。不仅考虑了设计的合理性因素,而且在细部的风格化处理手法上,也都减轻了建筑庞大的体量造成的沉重感和压抑感。

大厅是整个购物中心的核心部分,伸展的屋檐是gmp事务所惯用的手法。这是他们的独到之处,是一种对建造工程艺术的痴迷,他们知道如何将此运用到建筑的审美高度。克佩尼克购物中心的大厅宽18m,高25m,长150m,其体量显而易见是超常规的。大厅连接着主要通向巴恩赫弗斯达夫厅的主入口,后面是乌赫勒厅,并作为景观设计进行综合处理。乌赫勒厅面积不大,但优美别致。拱状的屋檐下玻璃、轻钢结构被拉力缆固定。数座桥跨越大厅,宽度大于常规,从而留出宽敞的空间可以安设冰淇淋小店。这种处理似乎更倾向于业主的行业需要,而并非设计师

初衷。建筑其余部分的内部设计按照街道景观进行:花岗石的地面,建筑外立面使用的红色石材在室内墙面和窗框处也屡屡出现。室内种植有树木,并用山毛榉木制作的长凳排列在两边。由于位于城市中心,因此在建筑四周没有设置停车场,但是顶层只有停车位。事务所项目负责人约基姆·林德将大厅内通向停车位的通道进行了巧妙的隐藏处理:通道外壁镶满镜子,顶棚采用玻璃材料。尽管大量运用美学的设计手法,但丝毫没有影响其愉悦功能,购物中心呈现出典型的都市广场的氛围。这就是个性实力的展现。这个建筑物打破了城市改排商店的常规,在三、四年内也无需再装修;此外,大厅的宽敞和开放性也是其特点。这两点也是它的委托商们争论不休,犹豫不决之处:他们不愿意将这样一个场所设计成半公共场所。这里不能有狭窄封闭的感觉,也不能允许大声喧哗。温度宜人,主要通过空调调节,当然也不能过热,使人们在冬季也可以身着外套在厅内行走。由于大厅很高,因此在热卖时也不允许挂广告旗帜。整洁合理的形式,坚固高质量的材料,这就是克佩尼克商业中心出奇制胜的信条。如果要勾画出从传统的百货商店到现代购物中心的转换草图,克佩尼克商业中心就是

一个超凡脱俗的样板。总而言之,与地下的杰桑布鲁纳中心以及未来建造在威定的杰桑布鲁纳城市特快站相比,克佩尼克商业中心明显占优势。克佩尼克和威定几乎是同时在1997年秋季开业的。

杰维·佐伦

折叠能够再延伸吗?

美学的碰撞:与霍恩的连接问题

弗兰克富特·阿尔杰敏
ZEITUNG报 1998年5月28日

基尔市位于霍恩河畔,城市最远处是霍恩。霍恩河一边是历史悠久的城区,另一边则是工业区和轮船码头。基尔市的这一边被人们视为"贫民区",大多数聚居者为工人阶层。正如德国其他地方一样,随着曾经繁荣昌盛的造船业走人低谷,这个地区也倍受冷落。德国造船场这个大型企业为城市的发展留下了大块土地,赋予其新的用途并将为新都市的开发提供良机。早在20世纪90年代,一个新的城市格局面水而立,基尔市建造了许多住宅,办公楼及休闲场所,并准备将一个斯堪的纳维亚式的摆渡点移到"狭长海湾"的另一边,一个新的工程就此启动了。惟一的先决条件是:霍恩的水路

要由一座桥连接城内主要的火车站及新摆渡点。然而这座桥必须能折叠,因为这里是狭长海湾的尽头,因此又是旅游船和当地摆渡船的停泊点。设计任务相当明确事务所总设计师奥托·弗拉格提出了一个惊人的设想:设计将由gmp事务所同期图塔噶特工程公司下属的斯罗德·本杰明合伙人公司共同完成,而这家公司因其精湛的工艺和创新构想而闻名

于世。强强携手,他们设计出了惊人之作:正是前言中所述的世界首创的会林柄钢缆开、合的"三节桥"。可以想像一座桥的结构如同不偶的胳膊,不同之处在于它的上下臂及手三部分是一样长的,它的开合机械类似折合的手臂,连接在它一边的肩部上。这个机械装置为船通行提供了水动力启合功能。桥的两部分可以由钢缆自如操纵,然而第三部分就增加了难度,这是显而易见的。因为既要保证夏季的操作无误,又要确保冬季能够承受入港的寒风、冰流及低温。

城市建筑权威及设计者们不仅考虑到这个项目为步行者与骑车人提供方便的功

性需求,而且更关注的是将其设计成为基尔市的标志,是向着新技术、新区域于敦的新起点。但是只有城市设计委员会欢迎这个革新,加顿区的居民并不以为然,并且提出反对意见。反对基督教民盟协会认为有义务出面拒绝此项目,计划委员会和海港指挥也一致反对此设计。社会民主党意见分歧,争论矛头不是指向桥本身,而是指向奥托·弗拉。然而最后的进程平息了所有的疑虑并解决了所有的建筑难题。当罗斯托克市贝佩顿码头正在进行测试时,桥梁设计的各项指标均告成功。那些表示疑虑的人也产生了动摇,甚至媒体也改变其原作为反对派的初衷。审计郑重声明:一座造价不高,独具特色的新桥梁提供通往在荷兰的通路。大标题"一座奢侈的桥"由"基础设施"所代替。当连接城市中心的诺言被推迟时,海湾的连接促使了工程的进展。由于当时正在进行地方选举,因此没有人对实验的进程和改进感兴趣。新上任市长诺贝特·甘塞尔快刀斩乱麻地作出决定:一个临时的桥花费大约100万马克,而正式的桥还在试验中,因为已经进行了大量的前期投资。然而6个月之后,计划提前了许多,而且比预算还省了30%的经费。
一座装点着红、黄颜色标志,仅仅花费大约350万马克的新桥建造起来。这座桥被认为是冲破重重阻力脱颖而出的巨大喜悦:像成年人的玩具,像延乡里流动的雕塑,但更有使用价值。当大量的船只通过的时候这一点就更加明确了。这座桥能给基尔市带来很大的收益:作为新港口的象征,它的发展会在今后几年中得到验证。同时这座桥也将一个未来城市区域与传统面貌而立、优雅美丽的住宅及休闲区连接了起来。作为一个高科技的建筑标志,它赢得了这次挑战并且增强了造船业方面的机遇。一个移动的缆绳拖吊拉动起相反作用的吊板,用分半钟的时间即可开启桥体,非常有实效。如果能够从最新的荷兰桥梁建筑而引发灵感的话,基尔市就能够骄傲的更早些了。在鹿特丹建造的横跨马斯河的新建桥梁(也被公众批评为超支)的城市议员获得了设计奖。评审团的意见是:在设计中,艺术价值高于商业价值。

杰特·卡尔勒

勒尔特火车站 柏林

(《建筑》1999年7、8月刊 第139页)

柏林最重要的火车站之一将要建造在洪堡特区东部,勒尔特车站旧址上。勒尔特车站的设计是gmp事务所的又一力作。gmp事务所的作品已经在两德统一后的建设大潮中尽显风采,独领风骚。他们的"设计"在世界各地留下了重要的印记。在这个国度里建筑业被认为是一种将工程技术智慧与城市肌理连接起来的结晶。这也构成了决定建筑形式及选材的两个关键的特征。gmp事务所的设计在无数基础结构的处理方面表现出色。他们的主要特点如下:首先也是最主要的是善于捕捉那些棘手的工程,其次材料的使用是如此之创新。例如:车站的四个铰链拱形框架,这就是设计中棘手的课题,因为它涉及到工程学及相关学科。在设计时他们先来看一下环境现状,或者换言之让我们关注建筑的尺度问题。总建筑面积164000m²,其中75000m²作为商业用途,4300m²用作铁路服务设施,19000m²为运输分流用途,剩余的35000m²将设计为车站站台。表面上看来火车站的设计无非就是几根简单的线条问题,但考虑到其整体布局分析,事实上需要煞费苦心地对空间进行三维立体布局。设计需要切实解决都市空间的膨胀以及包括内城服务设施安排的复杂的多功能空间的难题。车站的核心空间是一处两层高的建筑,此外,地下15m深处有隧道(供横跨德国东西方向的快速列车行驶)以及一条建在地下10m处南北走向的U—巴赫铁路线(另一条高速线路)。gmp事务所在设计这条通往复杂城市中心的国际线路中表现出极精湛的设计技巧,获得了极具冲击力的影响。

这些"巨型"项目与传统城市尺度相比有点"出格",通过精心的设计,他们试图将城市改造成连接国土的中枢车站。建筑师意识到了合理设计的重要意义,换言之,是利用技术与智慧发挥设计与规划的潜在可能。首先也是最重要的意识到建筑结构就是"身体"的骨架,骨架是人体保留时间最长而且最有形式感。建筑的形态特征以及人类空间的转换都不能对此视而不见。虚伪地承认建筑的前途在于对以往建筑的照搬模仿丝毫没有意义。未来的建筑师应该深入研究结构科学,利用计算机科学对建筑结构问题进行解决,为新设想的空间难题作出实质性突破。

要想遮住年轻人的眼睛,对他们讲述历史上陈俗的故事,盲目的继承是无用的,那种一味地模仿过去也不再是现实。古斯塔夫·亚历山大·艾弗尔的哲学历史学说教育我们:很早以前,生活已经在影响着技术的发展。

马里奥·安东尼奥·阿纳柏尔迪

宏伟的交通中枢

柏林最大的车站即将诞生，它的面世将成为连接南北、东西线路的中枢。欧洲铁路系统几乎是以拥抱的姿势来迎接这一伟大时刻的来临。

（勒尔特·巴赫霍夫，柏林）
建筑：gmp事务所
《建筑研究》1999年第1期）

勒尔特·巴赫霍夫车站将成为柏林主要的火车站。战前，在这里曾有过一个车站，但是已经遭到破坏。现在只剩下一个极大破旧的S-巴赫车站高高地卧在铁轨上。重建的工程已经在进行，其中包括两项连接欧洲大陆快速列车线路的计划：南北线路从斯堪的纳维亚至西里岛；东西线路从伦敦至莫斯科再至亚洲。

这里还要建一条新的南北走向的地铁（U-巴赫）以及一条东西走向的车站（S-巴赫）。这个车站是为整个首都建造的，当然其他的车站还在使用中（例如：动物园-巴赫霍夫专线）。勒尔特-巴赫霍夫专线将为行政区数千名公务人员与北部默阿布人聚居区提供交通要塞。这条铁路的建设将成为被遗弃多年的那部分城市的催化剂。诸如此类的城市中：施佩尔曾是北部纳粹的一个中转站，也是南北方向的中枢线路，1942年被英国空军炸毁。勒尔特车站的设计是大胆而又简捷的：东西线与S-巴赫线在同一平面，在街面以上10m处。南北线与U-巴赫线在地下15m处占据许多隧道中的两条隧道。位于城市中心之下，远离流砂层。（只要不影响柏林首都的地位，地下没有固定的限制）地下施工，同时需要考虑到地面还要为汽车、公交车、出租车、行人提供服务。地面的站台设

计非常壮观，由430m长的玻璃篷顶覆盖，屋顶和站台将穿过两座平行的条块状建筑，50m宽，170m长。要将其正好设计在地下线路的上方，仅与长车棚的中心弧线稍微有点偏斜。在两座平行的条块状建筑之间设有商店、服务设施和旅馆。多层的车站大厅又是被玻璃屋顶覆盖着，阳光可以直射入站台底层南北线的列车与U-巴赫线。条块状建筑和玻璃屋顶都采用特制的轻钢结构，钢板外部还有格子框架：是直截了当的、理性主义的、普鲁士式的特色（虽然建筑师们的办公室在汉堡）。屋顶是由设计精良的轻型壳状顶层结构覆盖着，菱形的方格是1.2m×1.2m，每一个单位都用斜缆拉紧，而所有的方格都由构架支撑着，据称它就如同一列滑铁卢格林萧的双层列车（1993年9月）。非常有趣的是看到这两种结构被紧密地结合在一起，这个车站的确能达到建筑师们预期的效果。勒尔特-巴赫霍夫车站被认为是进入城市的门户，因此要浓妆抹扮。车站大厅给人印象最深，它是最具有活力和激动人心的地方。当首席大法官从他的办公室向北看去，站台上方的玻璃屋顶给人一种富丽堂皇的视觉效果，它具有复活的双重意义：城市的复活和欧洲铁路系统的复活。

安娜·维纳

为未来博览会设计的四座塔楼

新博览会区计划申述：格雷戈特·艾戈瑞德设计
汉堡实际建筑地面面积95000m²，1998年始建
《IL RESTO DEL CARLINO》，1997年11月15日）

展览区可以分成数个展览空间，可独立使用。新博览会地址设在下萨克森州的策勒市。董事长勒伦佐·凯格诺尼声称：博览会新馆将在2000年秋季落成。在欧洲区域广泛的设计竞争之后，昨天gmp事务所从众多竞争对手中脱颖而出，此项工程将由久负盛名的gmp事务所完成。gmp事务所总部设在汉堡，因其设计的作品斯图加特及汉堡机场、莱比锡及汉诺威展览场地而闻名遐迩。此项设计以往没有任何雷同之处，与原设计不同，这个设计建议停车场设在屋顶区。在里米尼设计的博览会新馆好似传统的意大利建筑的照片。展区的建筑面积为95000m²，包括展区，技术服务区。建筑设施的构思是一层结构，主要入口处在南面，与当地的交通枢纽、火车站相连接从而成为里米尼-里西尼的纽带。在博览馆的东西处也设立了出口区，停车场可提供5500个车位，形成上与三个入口处瀑洽。四座用灯装饰的塔大约有30m高，在入口处将要表志。前面的场地扩展为一个开放式活动广场，在重大场合时灯光可覆盖整个场所。一个宽敞华丽的环型小大厅的直径为50m，上面覆盖拱顶，坐落在展区中部。博览馆行政管理办公室，会议中心，各种公共服务设施，餐饮均设置在此。展览大厅宽62m，采用无支柱大跨度式设计，充分利用地面空间（博览会的办公室和服务设区设在大型玻璃圆拱顶的周围）。进入每个展厅的两排柱廊自然地形成了入口处。展览区与入口处的设计依据各有不同的用途尺寸不一。设计还考虑到未来

馆将扩大到125000m²的建筑面积,囊括16个大厅。董事长勒伦佐·凯格诺尼声称设计方案将在3月份提交询问,由博览会董事局咨政审定通过。建筑将在直接招标之后,1998年秋季开工。投资预计1790亿里拉,1530亿用于建筑设施,其余的资金用于其他相关设施等,两年后投资大约1570亿里拉。成本高低不一(从0亿到130亿不等)是由于土地购买价格的不同。

变幻莫测如同时来运转;跌宕起伏如同赌场兴衰

今天是"坏小子"史特本游乐场的开业典礼;这是一个焕然一新的建筑。
《弗兰肯邮报》1999年10月日)

经过几年的准备及投建过程,在这个地区一个独特的建筑物"坏小子"史特本游乐场迎来了它的开业典礼。设计师是汉堡的迈因哈德·冯·格康,他被认为是德国建筑业的顶尖人物。

"坏小子"史特本游乐场

这对迈因哈德·冯·格康来说还是初次尝试。尽管这位著名的建筑师的作品遍布全球并赢得了百余次设计竞标,但这还是第一次设计这种游乐场。迈因哈德·冯·格康针对为什么要参与此项工程这个问题回答说:"对于建筑师而言,每一项特殊的设计工程都极具吸引力和重要性"。游乐场就是独具典型意义的一例,它能够给我们提供一次尝试和发展新建筑风格的机遇"。迈因哈德·冯·格康介绍了大量的寻求好运的赌场的参考设计及周围景色的设计。例如波状的钢板屋顶结构:既像赌博时运气的起伏跌宕,又像斯迪芬的泉水蜿蜒崎岖流过游乐场;玻璃的外墙;在某种程度上象征着这种材料也像好运一样容易破灭。波状的钢板屋顶自中世纪以来只在弗兰肯与图林根山区低矮的建筑上。建筑师鲁道夫·翁格劳布解释道,这座建筑物在工程实践方面将起到重要的作用。工程监理鲁道夫·翁格劳布认为游乐场的设计及功能是高质量高效率的,它将起到一个正面广告的效应。

迈因哈德·冯·格康与他的合伙人玛格一起创立了德国最成功的建筑事务所,拥有300多个职员。他的代表性设计包括卢卓克古典音乐厅、受到高度赞扬的勒尔特·巴赫霍夫火车站,以及莱比锡的博览会。在此他自己评论到:设计进行在现实

与理想的边缘地带。在柏林德国国会大厦重建项目的竞赛中,gmp事务所位于英国建筑师诺曼·福斯特之后名列第二。ZEITUNG日报还将迈因哈德·冯·格康列入一小部分世界知名建筑师行列。此外,名单里还包括他的伙伴福尔克温·玛格及加利福尼亚的弗兰克·欧·盖

里以及意大利的伦佐·皮亚诺、英国的理查德·罗杰斯等,当然还包括英国的诺曼·福斯特。格康于1935年出生于拉脱维亚首都里加,父母在战争中双双去世。1945年他来到汉堡,从此他便在此生活工作了。冯·格康曾同时在几所大学中任教,如:不伦瑞克大学、东京大学、比勒陀利亚大学,他不断在论文及书中谈论建筑理论,并用理论来说明自己的信念,精益求精,追求完美。建筑是一门社会实践的艺术,它创造人类的生活空间。"因此良好设计的标准就是不断追求的平凡的幻想曲。"

在众多的项目中哪一项是迈因哈德个人最满意的呢?对于这个问题的回答是公共设施对他来说是至关重要的,因为它们不仅要满足个体功能的需求而且还必须满足城市及社会的需求。诸如此类的建筑物如:汉堡和斯图加特机场,罗布特中心及柏林市中心车站。此外他还将自己的设计作品比作自己的孩子"作为自己的孩子,你很难说哪个是最心爱的。"
埃尔弗德·施奈德

不要惧怕蓝、黄和红褐色

色彩的学问在于色彩的铺垫:两种新的混合色的审视——奥尔登堡和开姆尼茨一改灰色背景。

(弗兰克福特·奥尔盖尼《ZEITUNG》1999年1月25日)

"不管用什么色彩,墙、圆柱及顶棚的颜色必须是灰色"这样的回答几乎已经成了建筑师们对业主的一致答复。但是对于汉堡的gmp事务所而言精益求精也体现在颜色的选择上。现在对于颜色的观念也在起着某些变化,大面积的墙面上涂着赭色坐落在黄色水泥的结构上令前来开姆尼茨技术学院参观的人大为吃惊。礼堂内部蓝、黄、红褐色交相呼应,色差对比较大,给人一种焕然一新的感觉。因为开姆尼茨技术学院位于城市的南边,坐落在高耸云天的塔楼之间,因此它们的建筑式样毫无疑问会给城市增添色彩,

但是它们的抽象安排格局并没有给城市空间提供什么色彩，只是在一天天的磨损。这可能就是1994年gmp事务所赢得开姆尼茨技术学院的重建与扩建项目竞标的主要原因。他们提出一个有力的全新的建筑设计，坐落在楼群之间，一个南北走向的广场成为一道新的风景。会堂中心与对面的规划的图书馆将形成学院的重点部位，起到连接主干道及风景区的作用。这座建筑与周围黯淡的色调形成了强烈的对比，独ську风骚。强烈的色彩安排都是为了一个目的：一

改传统的灰色。一排大学教室和一个大门廊在两座礼堂下面。从上面阶梯教室倾斜的阶梯向外望去，会给参观者一种宽阔的视界。低矮的建筑物外部棱角用镀着银灰色的波纹钢板覆盖着。礼堂外侧安装着钢制安全梯，它的安全设施也独具"技术含量"。中间部位突出色彩概念，阶梯教室部位呈现艳丽的黄色，高高悬挂在两层楼的入口大厅上，而大厅的下层漆成了红色，上层却漆成了深蓝色。有时它给人一种抽象派的印象，因为颜色不是分散到每一座建筑物

上，而是强调了不同墙壁及屋顶部位，给人强烈的兴趣。但这并不是千篇一律的：当建筑材料的传统概念与这种创新思维使用颜色相碰撞时（在黄色墙面前面的就使用了浅色木扶手）这便成了画龙点睛之作。但是，总体来说这种概念是正确的，因为建筑物的立体感是由颜色衬托出来而产生的一种独特空间立体。

6个月之前在奥尔登堡由同一个建筑师完成了又一相同的设计任务。尽管它的建筑风格完全不同，但奥尔登堡大学却被视为色彩效果的实验地，因为它的休息室和局部立面有一种诙谐的效果。在其他方面还增添了木材、钢material的自然色以及色彩艳丽的水泥流行色。

除此之外，与这两座会堂无法比拟的是：在奥尔登堡的卡尔·冯·奥西兹库大学，这是惟一的主会堂，它被安排设计成一个与其他具有地方色彩建筑群体相连接的主体建筑。致使这座建筑物具有了与其他建筑物完全不同的建筑理念。圆柱体的建筑结构位于城市的西部，濒临闹市区，而且意识到学院本身就具有各种不同的功能。墙壁延伸与缩进处，开放的休息厅的一边多变化的立面设计与主楼关闭的演讲厅已经将建筑物立面纯粹的几何图形打破，一条长廊一直延伸至演讲大厅，圆柱体

的建筑物当中。

纯几何图形抽象概念的运用：圆柱体、矩形装点着白色石膏材料的立面，只有一面涂着淡褚色，就像一个

白色背景下的舞台布景。休息厅内部深蓝色，但在木材本色、地板及水泥圆柱的衬托下减轻了它的对比度。

gmp事务所完善了这一新的设计，发现了颜色的效用。这一发现可以促进许多领域的发展。总而言之汉堡设计室正在忙于各种大型的设计（勒尔特·巴赫霍夫的奥林匹克体育馆的翻新工程；新上马的舍内芬德飞机场），它们将在未来向柏林以至全德国的公众显示其独特风貌。自然材料的高贵限制是有其自身的局限性的，色彩的作用增添了它的额外成分，同时也扩展了建筑业的视野。

格特·卡勒

城市的活力—旧"欧洲"艺术博物馆的翻新

艺术收藏家、投资人汉斯·格罗思计划在"韦瑟班霍夫二世"原址上修建一座艺术

博物馆和服务中心。

（A&W，1998年第11期第70页）

"韦瑟班霍夫二世"博物馆是集古典与现代特性为一体的建筑。对于这座建筑物的可行性设计是由汉斯·gmp事务所进行的。他们将在"韦瑟班霍夫二世"的原址上修建一座艺术博物馆，收藏德国的画家和雕刻家安贤基尔夫收藏艺术品。这座博物馆是由艺术收藏家和投资人汉斯·格罗思发起并出资修建的。他为在这个原址上建筑艺术博物馆、办公室与服务中心拉开了序幕。

未来之舟的终点站？

古典欧洲艺术博物馆的都市化重建是从一个文化建筑开始的。当"蒂霍夫"还在文化发展的配酿之时，艺术收藏家和投资人汉斯·格罗思对其1996年购买的这块16500m²的土地用途已经有了明确的意图和

向。当420m²的博物馆建完之后，一个占地30000m²的服务中心将作为二期工程投资建设。由于最近又添加了办公空间，这个商业建筑被作为中期项目开发建设对于社区内基础设施的建

也将在相应的时间阶段
开。将劳埃德里广场扩展
威悉河已经被列入计划，
包括一条位于街道和铁
下面供电车、自行车、行
使用的通道。这个新的通
将连接城市中心和建立
悉往返列车落脚点的"白
标线"。作为一个旅客中
站也是可行的。

板立面材料

巴西里佳艺术博物馆的
划已经完成，将专门为安
·基尔夫新的艺术品提供
室。殿堂的中部是高9m，
8m，长20m的展览大厅，
过顶棚的光槽照明。狭窄
走廊倾斜的屋顶表面有限
光泽表现出一种庄重之
。立面上采用的铅板材质
强化了这种庄重感。格罗
解释道：如果安贤·基尔
考虑将目前在马德里展示
艺术品在此展示的话，本
馆将在1998年秋季开工。如
第一批展示的艺术品没有
过格罗思，"此项工程将
推迟一年"。总之，这个艺
馆的设计必须与展出的艺
品相匹配。

兰德海尔姆·菲尔德霍斯

教学与研究建筑
RESEARCH AND TEACHING

开姆尼茨技术大学阶梯教室

竞赛: 一等奖, 1994
设计: Meinhard v. Gerkan
项目负责人: Dirk Heller, Astrid Lapp
合作者: Angelika Juppien, Knut Maass, Ralf Schmitz, Michele Watenphul
业主: Staatshochbauamt Chemnitz
建造时间: 1996-1998
建筑面积: 8.856m²
建筑体积: 51.766m³

在两德统一后所带来的那段最初的自由及扩展的快感中,开姆尼茨技术学院为了雄心勃勃地扩大校园举行了一次城镇规划设计竞赛,首先要建一座大礼堂和研究楼,然后建一座图书馆。迈因哈德·冯·格康最后获胜后,在规划的构思中提出了一个线形结构,此结构具有长期发展的灵活性并且包括大礼堂和图书馆的入口广场。此项目是在所有目标项目建设中惟一将要投入施工的,并且它在原来的面积上进行了相当规模的缩小。幸福的快感随着发展空间渐渐让位于无限扩张的商业场所而化为泡影。现存的礼堂成为最后的结局而不是按最初的设想作为扩建一期工程构思的,这一事实反映在这一大型建筑

1 大面积粗灰泥墙面满足节约开支的需求。参观吸取墨西哥和巴拉干建筑的风格以及莱格罗特带来的色彩概念的启发。
2 城市背景中的建筑规划总平面图,包括技术学院扩建工程。
3 设计方案与环境现状鸟瞰图。
4 大礼堂的支撑墙体构造一个入口前厅。

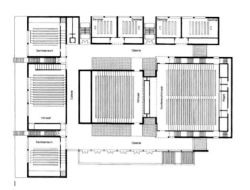

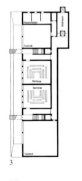

1 二层平面图
2 一层平面图
3 地下一层平面图
4 东面主要入口，墙体支撑着大礼堂界定出建筑的立面

结构上。由倾斜的走廊连接的位于两个大礼堂之下的一排研讨会议室构成了门厅的轮廓，其中设计了从底部也可以看到的逐渐升高的坐位。尽管礼堂的内部非常深，但建筑的布置使自然光线可以进入内部的大厅，并且在那里的垂直空间准备了顶灯，这样，则提供了一个明亮愉快的环境，与往常所想像的许多大学里的阴暗礼堂形成了鲜明对比。两个大礼堂也不属于那种所谓的"黑盒子"从空中的全景为学生提供精神上的刺激。另一个主要的设计特点是墙壁和顶棚的强烈的色彩。这与最初的设计概念并不相符，但这却是在认识阶段由于基本的经费限制和建筑师在墨西哥一行所产生的感想中得到的发展。大面积的墙面和顶棚表面的经济材料大多是粗糙灰泥，此后在灰泥上涂颜色也很合理。从巴拉干和莱格罗特建筑风格的参观中产生了运用简单的元素提出杰出设计的灵感，对颜色的大胆使用也是对其他城市里高大建筑灰暗单调

的反差。在减缩及细节方面也体现了这个建筑的朴实、简洁的原则。颜色的选择和房间设计采用平面取代空间的设想，完全是个增强实践过程的结果，并且开辟了迈因哈德·冯·格朵表设计理念的新篇章。经济上朴实的原则使此建筑的开支远远低于原来的财政预算，这在其他所有设计方案中都有体现，如：地板使用工业水泥砂浆、顶棚使用打空金属板、建筑的立面使用波纹钢板，此外在建筑室外还建造有一个钢结构的逃生楼梯。

2

1-3 室外逃生楼梯建于礼堂凹处之间而且直通外界

3

1 靠近二楼门厅延伸出的是一个宽敞的朝南露台,通过几阶楼梯可下到底楼。建筑的立面是经过充分考虑有限的经费而设计的。粗糙灰色水泥和波纹铝板在价值衡量方面是不可比拟的
2 两个礼堂之间北面的安全梯
3 从门厅可以对南面露台下面一楼的风景一览无余

下页:
一个清晰的体块组合划分礼堂和人流区,同时也界定了建筑的垂直交通

2

3

1 上层南面的门厅。天光与孔隙的设计保证了阳光进入建筑的深处。吊顶材料是网状钢板,铺地材料是水泥砂浆

2 纵剖面图

3 横剖面图

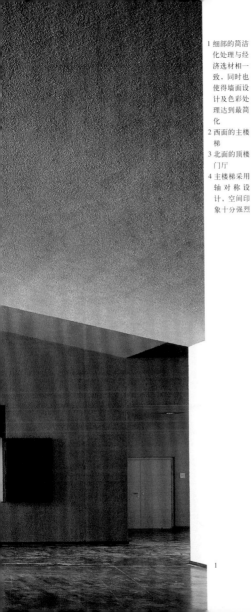

1. 细部的简洁化处理与经济选材相一致，同时也使得墙面设计及色彩处理达到最简化
2. 西面的主楼梯
3. 北面的顶楼门厅
4. 主楼梯采用轴对称设计，空间印象十分强烈

1 + 2 拥有1000个座位的大礼堂,天光设计提供了最理想的自然光并且光线还可以调暗

2

1 通往礼堂上层的楼梯,内设独立式坐席区
2 座位与墙面的枫木护壁

名为"开姆尼茨半导体"的互动式的雕像是由斯蒂芬·范休尼设计制作,用以强化楼梯的轴对称。

奥尔登堡大学阶梯教室

竞赛: 一等奖，1990/1992
设计: Meinhard v. Gerkan
项目负责人: Klaus Lenz
合作者: Karl-Heinz Behrendt, Bettina Groß, Susan Krause, Bernd Kottsieper, Jens Reichert, Dagmar Winter
业主: Staatshochbauamt Oldenburg
建造时间: 1996-1998
建筑面积: 5.163m²
建筑体积: 29.135m³

卡尔文奥塞斯基大学的主演讲厅建于一个以小型建筑区和20世纪60年代城市街区为特征的地区中的一条繁华街道的交叉口。这个演讲厅大楼可供众多教职员工使用，采用了两层圆柱体结构，气势雄伟，

1 大礼堂的"舞台一侧"，可划分为三个部分
2 鸟瞰此建筑可以清晰的观察到它的几何感和层次感
3 礼堂窗户外的遮阳装置的细部处理

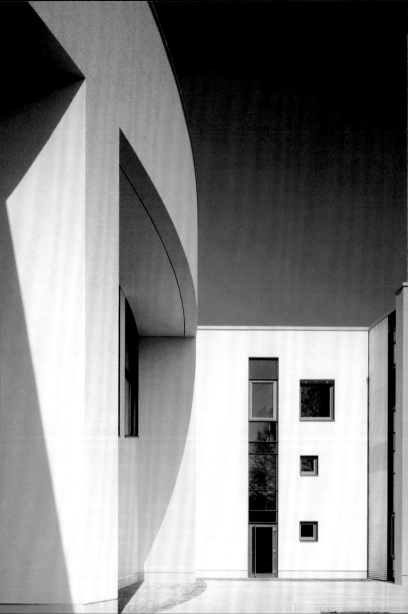

相互交错的圆柱形空间
一层平面图
二层平面图

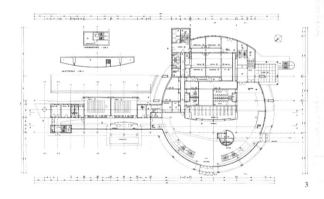

其与众不同之势也就彰明较著。突出与缩进的体块处理打破了纯粹的几何形体，并且通过前厅和舞台所在一侧在建筑立面的多样化美感的处理中得以强化。另外还有一座天桥连接着这个圆柱体直到学校建筑的出口，桥的终点是一座纵向建成的有演讲厅组成的建筑。白色灰泥的立面材料与建筑清晰的几何形状相辅相成。建筑外立面有两块地方涂刷色彩进行强调，而且在白色背景中俨然成为一道色彩风景。在建筑室内，一个废弃不用的电梯被改造成前厅内的一个自由体块，电梯被涂上了充满活力的蓝色，这与诸如木材、沥青地板和白水泥柱子等天然材质形成了鲜明对比。二楼的多功能厅配备了供戏院、会议和文化活动所需的技术设备，当几个活动同时举行时，这个多功能厅可以被分隔成三个部分。

建筑的构造

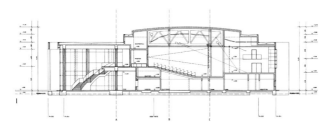

1 前厅和大礼堂剖面图
2 两层高的前厅室内
3 前厅内用材简洁：水泥砂浆地面和粗糙的灰泥墙面

1 环形与直线的几何形体，自然材料和有限的色彩强调构成了独特的建筑语汇
2 衣帽间弧线形的墙面采用木板贴面结构

1 舞台顶上的回声装置
2 榉木材质的座椅
3 透过讲台上方具有滤光作用
  的全景窗看到的室外风景

1

2

3

莱比锡大学，人文学院

竞赛：1997
设计：Meinhard v. Gerkan mit Stephan Rewolle

新的学院大楼将坐落于学校图书馆和迪米托博物馆旁边，并且与环境现状中一条所谓世纪之交的街道和城市平面结构形

3

态融为一体。选址考虑到了新楼的高度和街区的轮点，但仍然延续了图书馆的轴对称设计。结构上的轮廓包含了办公楼牢固的"外壳"和可灵活使用的"柔软"部分，此外，还包括其他空间的核心。底层的一条走廊形成了主要的纵向轴线并将内部的庭院面向城市开放。

1 轴测图
2 标准层平面图
3 总平面图
4 南立面图

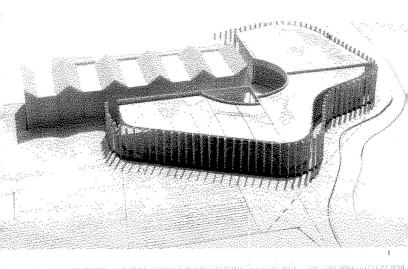

1

2

模型

剖面图

高级技工学校食堂
累根斯堡

竞赛：1998
设计：Meinhard v. Gerkan mit Stephan Rewolle

新食堂的选址是在一个主干道旁边，整个区位环境没有太大特色。设计构思主要体现在两部分建筑体块的对话中：一个是侧翼建筑，用作备餐的服务设施用房；另一个是食堂大厅部分，已经深入到周边环境之中。尽管餐厅内设有很多桌子，但空间的开放式设计特色仍然洋溢着一种欢迎的氛围。

1+2 模型
3 功能空间的分布
4 总平面图

柏林—阿德勒斯霍夫物理学院

竞赛：二等奖，1998
设计：Meinhard v. Gerkan mit
Nona Kazemi

就像分子结构一样，各种各样的正方体单位构成了一个大型正方体布局，正方体的大小是根据其功能的要求决定的。平面布局以一个整体的形式呈现于基地之上。一个内部环绕设计连接着数个分割的部分，并且使中间开放空间的正方体相对完整。这一构思将平面结构形态与明确的功能分区完美地结合在一起。

3

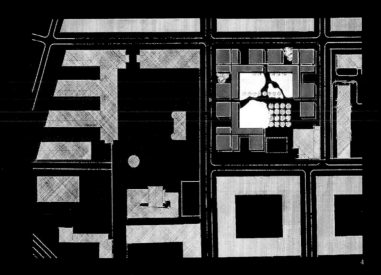

4

3

1 西南立面图
2 一层平面图
3 剖立面图
4 二层平面图

4

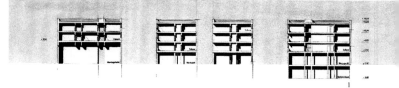

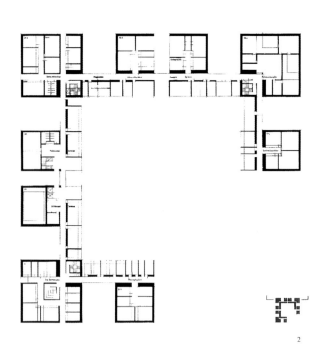

1 剖面图
2 三层平面图
3 东北立面图
4 四层平面图

地方体操馆 圣阿弗拉梅森

竞赛: 1997
设计: Meinhard v. Gerkan mit Nicolai Pix

建筑在基地上的平面形态极富变化,古老的树木和大片的绿地使之格外独特,建筑与自然景观从容对话。建筑物的外部形状大致相同,各单元均有一个中心公共区,四周是各种设施用房。设计建议将建筑建于即将拆除的旧楼原址,因此原有的绿地将被保护,被时光遗弃的旧址将被重新演绎。

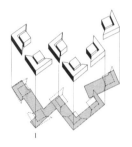

1

2

3

1 住宅楼布局
2 总平面图
3 中等体量的住宅楼
  二层平面图
  三层平面图
  一层平面图
4. 横剖面图

4

体操馆　柏林

竞赛:四等奖,1997
设计: Meinhard v. Gerkan
合作者: Michael Biwer, Nicolai Pix

这些建筑完成了原有的城市开发,采用U字形布局,环绕着学校的校园。进入到建筑内部,标准教室环绕着专业教室和辅助教学区,同样的U字形布局被重复使用。

由于各单元被宽阔的走廊分开,它们在"学校建筑区".总平面布局中清晰可见。主会议厅可以从里面或外面进入,因此它可被用于公众活动,毗邻的音乐厅可为公众活动提供临时的舞台。

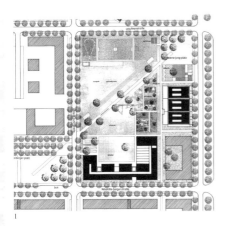

1 总平面图
2 顶层平面图
3 南立面图

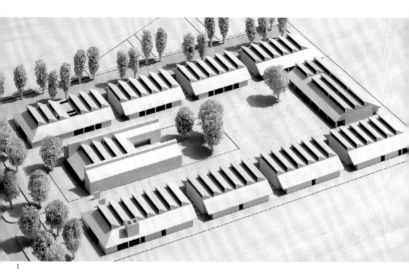

教学楼二期工程 柏林—马扎赫

竞赛：一等奖，1997
设计：Meinhard v. Gerkan mit Johann v. Mansberg

将高高的坡屋顶中心大厅和周围较低的附楼融为一体，这种建筑形式既简洁又令人印象深刻。这就意味着创造的是一个不受市区周围毫无特征的商业区和工业区影响的区域。因此，整个区域被看成一个"内部庭院"：一个商店区也可被用作送货区和课间休息区。这里也是训练基地的其他活动中心。

1 模型
2 总平面图
3 东南立面图
4+5 建筑及屋顶结构细部
6 一层平面图
7 东北立面图

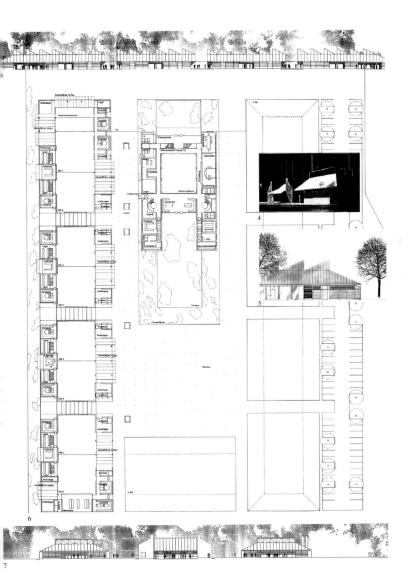

德语学校及职工住宅 北京

竞赛: 一等奖, 1998
设计: Meinhard v. Gerkan mit Michael Biwer
合作者: Bettina Groß, Elke Hoffmeister
建造指导:
Klaus Staratzke, Michael Biwer,

Sibylle Kramer
合作者: Michèle Watenphul, Knut Maass, Uli Rösler, Diana Heinrich, Jörn Bohlmann, Rüdiger v. Helmolt
业主: Bundesrepublik Deutschland, BRR
建造时间: 1999-2000
建筑面积: Schule 9.658m²
Dienstwohnungen 9.657m²
建筑体积: Schule 33.534m³

Dienstwohnungen 30.923m³

在中国建造一所德国学校本身就是为其国家做一个实在的广告。尽管如此,建筑在设计方面依然要满足基地的种种限制:传统的围合理念要求,复杂的建筑清除规定以及地边界的建造限制等。

目要求有一个100m跑道的运动场,在具体设计发展成为一个功能"复杂"的设计理念:这座简单的建筑包含教室、托儿所、体育馆,此外还有不同

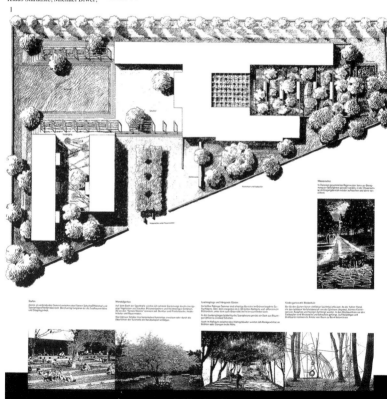

它的公共空间及体育馆屋顶的"绿色中心"。两座9层楼的高层建筑和低矮的教学楼构成了一个围合的整体。这个建筑充分实践了在一个拥有2000万人口的首都城市寸土寸金之地的综合利用。

1 景观规划
2 竞赛模型
3 总平面图
扩建设计概念
入口立面图
建筑立面细部

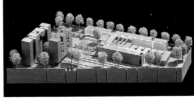

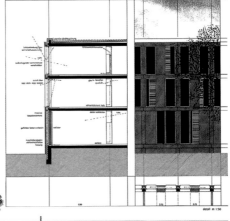

63

1 一层平面图
  功能多样化
  教学楼纵剖面图
2 体育馆上方的支撑结构体系
3 二层平面图
  体育馆内视图
  体育馆横剖面图
  大厅横剖面图

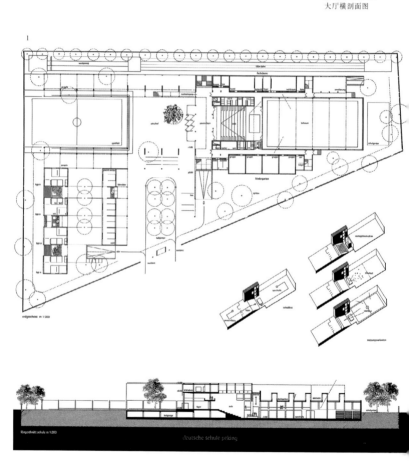

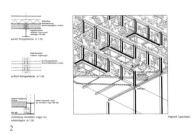

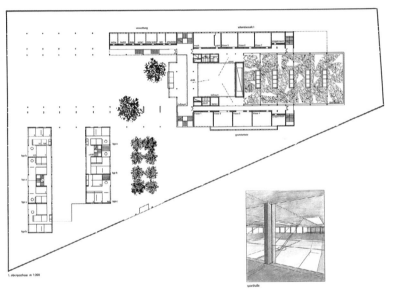

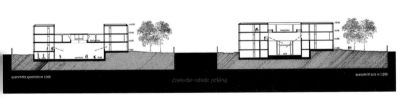

1 教学楼地下室平面图
  体育馆上的屋顶露台
  大厅内视图
  北立面图
2 三层平面图
  走廊
  南立、剖面图

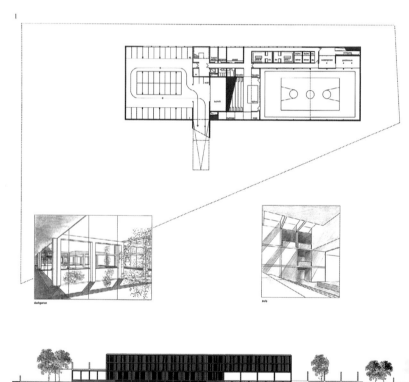

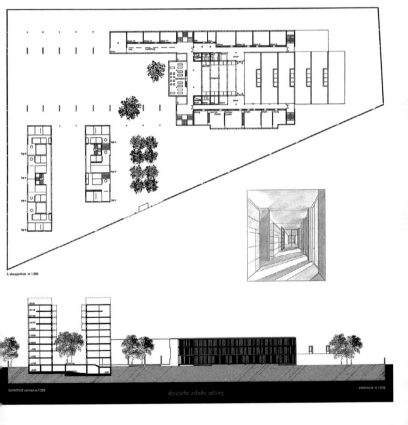

北京

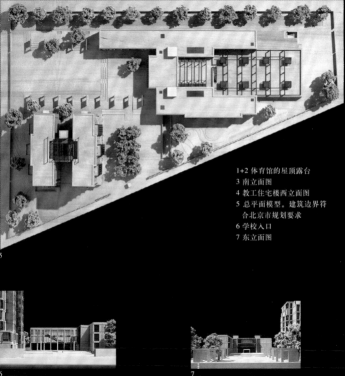

1+2 体育馆的屋顶露台
3 南立面图
4 教工住宅楼西立面图
5 总平面模型。建筑边界符合北京市规划要求
6 学校入口
7 东立面图

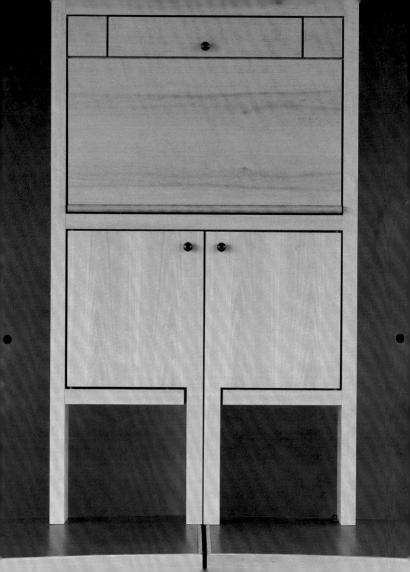

演讲台

设计: Meinhard v. Grkan, 1990
建造: 1998
合作者: Volkmar Sievers

当平直的靠在墙边时,这套折叠家具显得特别朴实,但当它的腿折出来桌面抬起来时,它就显示出其惊人的个性设计。这套家具有着稍微倾斜的顶可以作为标准的演讲台,它还装有灯光和一个比较大的隔底盘,可以放水杯或其他东西。演讲台采用瑞士梨木制成,它还装有清晰可见的镀锌金属接头。设计逻辑与折叠时和非折叠时的功能逻辑相辅相成。

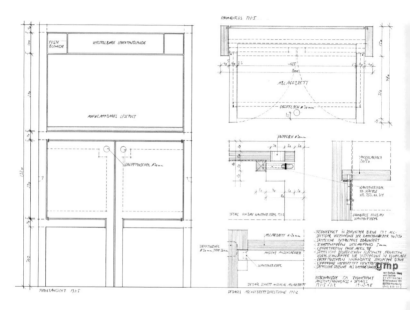

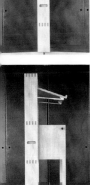
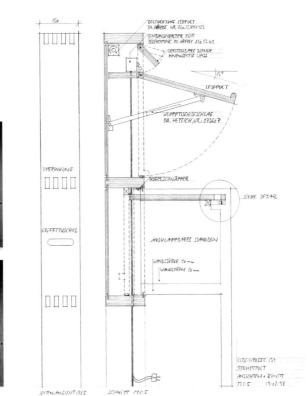

文化建筑
CULTURE

汉诺威2000年世界博览会展示厅

竞赛：一等奖，1997
设计：Meinhard v. Gerkan, Joachim Zais
合作者：Gregor Hoheisel, Sona Kazemi, Stephan Rewolle
建造指导：Joachim Zais, Jörn Orthmann
合作者：Ulf Düsterhöft, Matias Otto, Olaf Schlüter, Horst-Werner Warias, Andreas Hahn, Thomas Dreusicke, Helge Reimer
展示设计：Monika van Vught, Magdalene Weiß
委托人：Evangelisches Büro für die Weltausstellung Expo 2000
建造时间：1999-2000

建筑面积：2.004m²
建筑体积：18.548m³

为2000年世界博览会建造的基督厅综合了天主教和新教教堂的特点，设计意图通过一些建筑上的特色将这里营造成一个远离浮华名利的冥思之地，因此结构简单，简化到有限的材料，细部处理的精确细致，以及外观和空间氛围的别致独特。基督厅在建筑上清楚地展示了组合结构设计和细部处理。

整个设计选材保守而又简洁，主要采用钢、玻璃和水泥砂浆，此外，一棵古老的大树也成为建筑的点缀。一条3.60m宽，7.20m高的环绕回廊界定出整个建筑的框架同时也兼被用作展览空间。北面回廊包含了一个21m宽，18m高的正方体大厅，大厅的顶部用了9根细长的十字形钢柱支撑。灯光和强烈的垂直感突出了大厅的庄严肃穆，大厅可以从主广场面

1 概念草稿
2 构造模型
3 概念模型中的屋顶平面

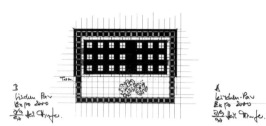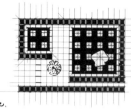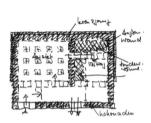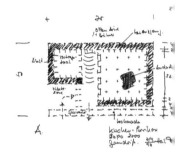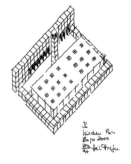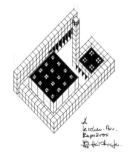

竞赛设计中的方案草图

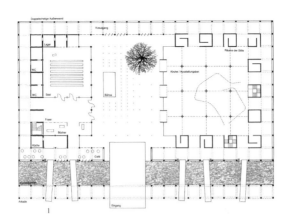

1 一层平面图
2 广场立面图
3 教堂立面图
4 纵剖面图

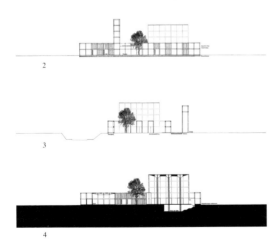

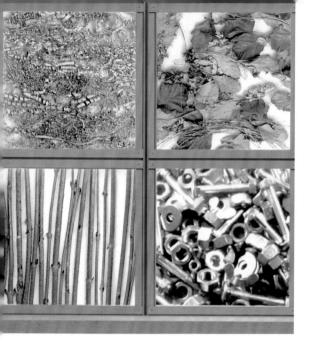
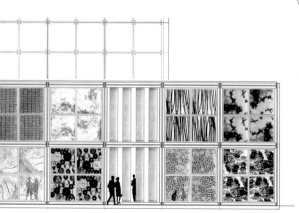
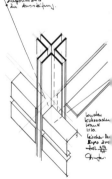

回廊的构造。3.6m × 3.6m的钢框玻璃窗内展示了各种各样的材料,充分诠释了"自然"与"科技"的主题。而汉诺威2000年世界博览会的整体主题就是"人类,自然,科技"。

环绕回廊的许多连接处直接进入。3.6m高的又深又宽的空间的"孤立地区"作为"沉默之屋"位于"基督大厅"和"回廊"之间,在那里基督信仰和教堂的主题通过语义学的解释传达给了参观者。楼梯连接着"教堂地下室",墙壁由白水泥随意勾画而出,3根横穿大厅的钢柱从广阔的屋顶向外继续延伸。光线的变化营造出这里的空间氛围。"基督厅"的光线是从装在柱子顶端正上方的顶灯发出来的,这强烈地突出了细钢柱的垂直感。薄薄的大理石表面构成一个光线可以通透的外壳,它鲜亮的色彩营造出另外一种空间氛围。相反,灯光照明强调的是"教堂地下室"的神圣感;暗淡的光线沿着柱子的轴线延伸,地板上环绕的条状的光线形成了一个焦点既突出了水泥材料特点又通过倒影的效果展现了一个神秘的明亮感。环绕的"回廊"装饰着双层玻璃立面,它被用来作为大型玻璃陈列柜。双层玻璃之间填充着各种来自自然和科技的材料,作为整体展示的一部分。玻璃墙根据各自填充物的不同,或多或少的成半透明状或部分地透明状。在汉诺威2000世界展览会结束后,整个展厅将要拆散并要在沃肯罗达和色林吉亚的展览会中重新建立并装饰成修道院。

1

2

1 教堂塔楼
 十字形
 标志性柱杆
 轻钢、玻璃结构的25m高的
 塔楼是2000年世界博览会
 的标志
2 "基督厅"
 9根细长的18m高的十字形
 钢柱成为这个庄严肃穆的大
 厅的特色，大厅墙面采用透
 光大理石材料
3 竞赛模型

3

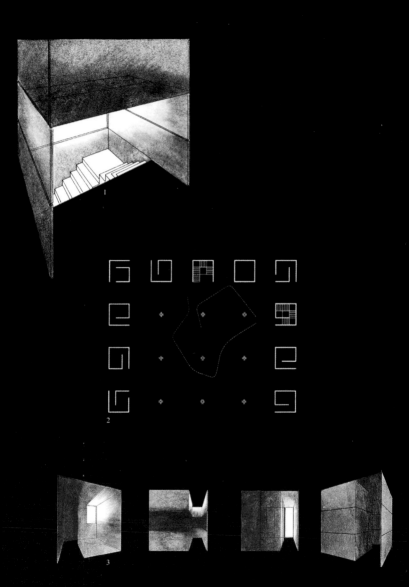

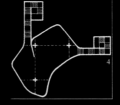

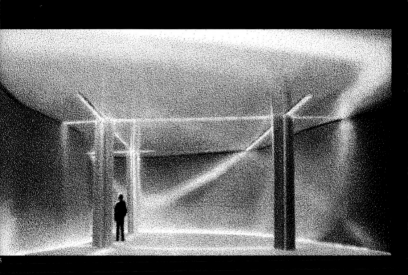

1-3 "沉默之屋"围绕"回廊"与"基督厅",墙面采用低合金高强度钢材料

4-5 教堂地下室,墙面采用水泥材料。上面基督厅内3根十字形钢柱穿过地下室顶棚落在地下室内。通透而下的光线强化了轴线感

最新规划——指日可待

世界展览会广场在建址上受到一定的限制，紧缩的预算也缩小了项目规划，但设计的构思没有受到影响。
1 从院子里看教堂的立面图
2 从院子里看服务区的立面图
3 教堂纵剖面图
4 一层平面图，从广场方向看去的剖、立面图
5 地下室平面图与北立面图

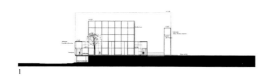

1

2

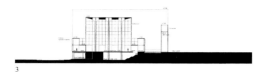

3

4

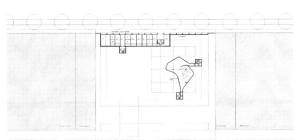

5

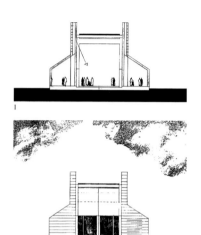

1 剖面图
2 立面图
3 平面图
4 模型正立面
5 总平面图

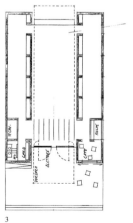

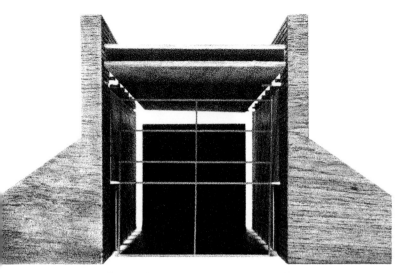

"韦瑟班霍夫二世"艺术馆

竞赛:一等奖,1997
设计:Meinhard v. Gerkan
合作者:Johann v. Mansberg, Stephan Rewolle

艺术馆新馆成为建在"韦瑟班霍夫二世"艺术馆原址之上的新城市街区的中心。建筑采用长方形廊柱大厅,整个展厅高9m,长20m,宽8m,线形天光照明。狭窄的走廊设计得非常隐蔽,与展厅的空旷形成鲜明反差。当代艺术家将在这个富有变化的空间内向公众展示他们的作品。

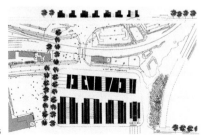

康斯坦丁 布宜诺斯艾利斯博物馆

竞赛: 1997
设计: Meinhard v. Gerkan
合作者: Sigrid Müller, Michael Biwer, Stefanie Driessen, Anja Knobloch-Meding, Nicolai Pix

一个艺术博物馆就建筑本身而言在城市里具有独特的标志性,它的设计效果在于为展品提供最佳的光线条件,这个设计同时满足了这两种需求。建筑采用双层玻璃立面,以弧形曲线穿过屋顶,带有层次感的天窗仍将南面射入的阳光进行了过滤。与此同时,这种立面处理也强化了博物馆的形式感。双层玻璃立面还围合出一个3层高的大厅,大厅通过电梯与相应的展厅相连,展厅层东面面对公园,展览在公园的风景中延续。

1+2 工作模型
3 总平面图
4 博物馆建筑的设计演变

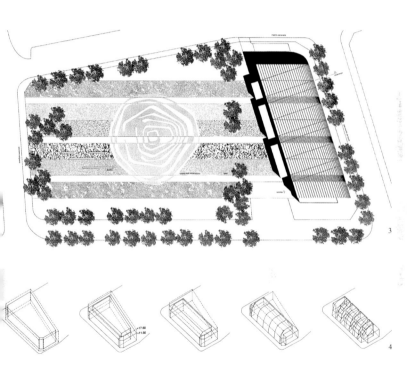

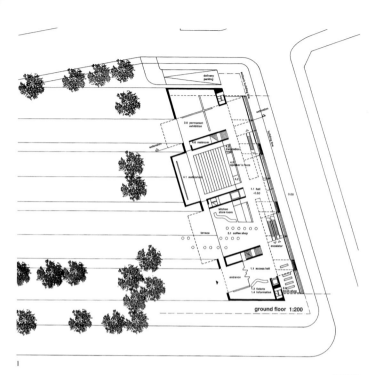

1 一层平面图
2 其他楼层平面图
3 横剖面图，立面图

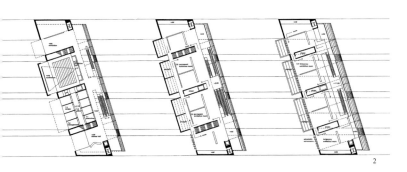

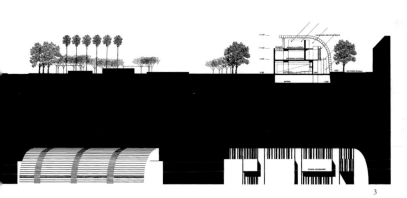

开姆尼兹工业博物馆

竞赛: 1997
设计: Meinhard v. Gerkan mit Philipp Kamps

博物馆建在已经废弃不用的工业建筑原址上，其扩建工程与博物馆融为一体，从而营造了新旧建筑对话的可能。

设计与基地现状平面结构布局形态相吻合，城区开发结构框架层次逐渐形成，而不是孤立地开发某一个建筑单体。灵活变通和历经多个工期之后的实施程度等方面决定了设计的最初概念的优劣，而这一概念当初只是建立在建筑原理之上。

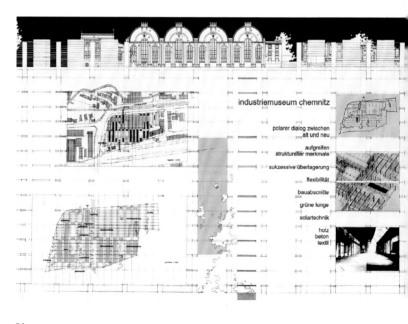

基督教社区中心 约翰萨林 汉堡

竞赛: 1997
设计: Meinhard v. Gerkan mit Philipp Kamps

基督教社区中心扩建工程将采用祈祷室的形式，扩建计划选址在原址后面，位于汉堡—汉夫斯特胡德传统的城市别墅之间。设计构思追随向周围地区延伸的不规则边界。有关鞍式屋顶的构思设计出一个"分割空间形式"的空间，连接并且通向了神坛的中心。建筑构造方式采用木材结构。

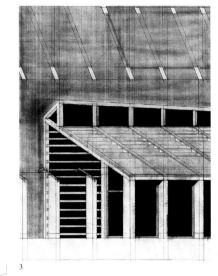

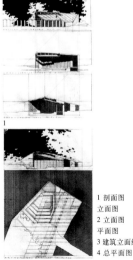

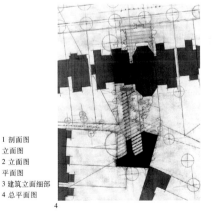

1 剖面图
  立面图
2 立面图
  平面图
3 建筑立面细部
4 总平面图

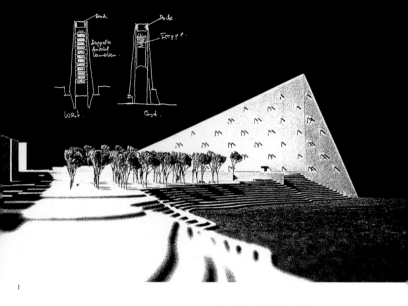

1

纪念碑 达喀尔 塞内加尔

竞赛: 1997
设计: Meinhard v. Gerkan mit Michael Biwer

纪念碑是纪念被贩卖为奴隶的非洲人的一个标志,不仅因为它根植于大陆的历史中,还因为它表达了一个对繁荣和美好向往的憧憬。纪念碑是一个令人震撼的个体,这个文化建筑的房间均位于底座层,同时也成为受压迫的标志,表达了受迫害和躲匿,还带有一种依恋非洲土地的感情。纪念碑从地基拔地而起:一种戏剧性的上升,狭窄的大厅宣告着过去的回忆。它的外部的象征意义有些矛盾性:纪念碑暴露在阳光下,一个"通往天堂的大门"连接着一座提供了横看大西洋广阔视野的平台。

4 模型研究与初步设计草图
总平面图
设计概念的隐喻

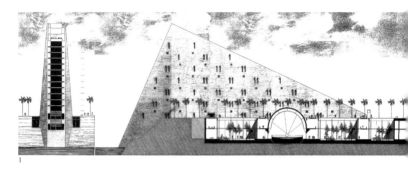

1 正立面图，部剖面图，侧立面图
2 横剖面图
3 纵剖面图
4 基座层平面

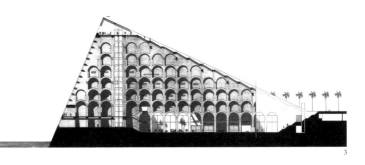

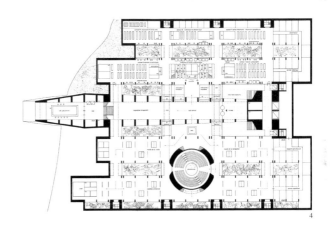

音乐厅 格拉茨 奥地利

竞赛: 1998
设计: Meinhard v. Gerkan mit Sona Kazemi

这是一个典型的几何布局：建筑沿着街道与基地之间的边界线横向延伸，中间部分在高度和长度上加以平衡，从而使中心空间成为一个横向轴线。建筑分为两个部分，与空间使用的要求相一致，因为一部分空间需要用作公共空间，而另一部分则仅供内部使用。建筑的核心是一个大型多功能厅，它连接着两个侧翼并将其向南面的麦兰皇宫和停车房延伸，那里设有公共入口。

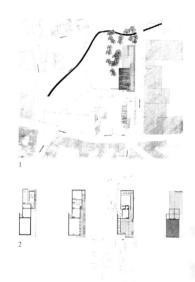

1

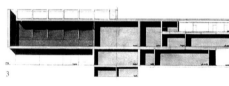

2

3

4

1 总平面图
2 三层平面图
　二层平面图
　一层平面图
　地下层平面图
3 纵剖面图
4 模型

社区中心 藻尔高

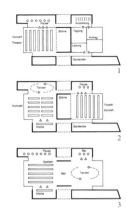

竞赛: 1997
设计: Meinhard v. Gerkan
合作者: Anja Knobloch-Meding, Sigrid Müller

这个新型社区中心拥有各项文化、社交和教育设施,在周边一带独具特色,它在林登大街地区的重要性是显而易见的。建筑的构造可分为一个主要的中央地区和两个外部的狭窄带状建筑,它们交错形成了面向林登大街的一个开阔的新广场,并且连接着休息厅。中心区域的新大厅使用灵活性强,而且面向城市以及户外风景。

1-3 社区中心的多种用途
4 北立面图
5 纵剖面图
6 总平面图
7 模型

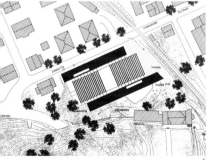

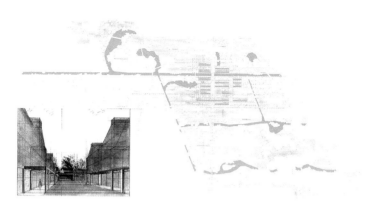

杜鹃花厅 不来梅

竞赛: 1998
设计: Meinhard v. Gerkan
合作者: Philipp Kamps, Pinar Gönul-Cinar

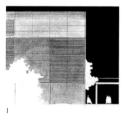

杜鹃花小路形成了连接中轴路和公园的主干道,它还成为通向各个温室的通道。在公园中,温室大小不一,成片分布,连起来呈现出连续而严密的平面结构形态。尽管建筑本身设计简洁缺少变化,但是由于沿着主轴线呈带状分布,从而达到了移步换景的视觉效果。建筑采用构造简洁的钢架和玻璃结构,并添加了可调节的百叶,对建筑的历史性风格进行了重新诠释。

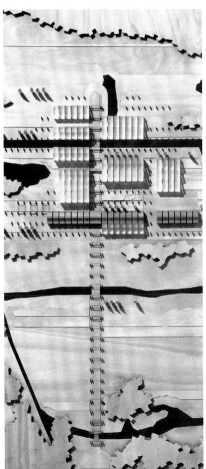

1 局部立面图
2 细部剖面图
3 大厅内的温度控制
4+5 结构模型

行政与办公建筑
ADMINISTRATION AND WORK

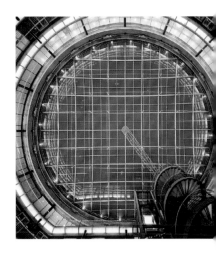

1 正在重建过程中的位于柏林巴林广场的大厅
2 总平面图 防火墙构成基地的纵向边界
3 垂直向可折叠百叶窗，它像可以眨动的眼睑一样为建筑添加了一个有特色的外表

德勒斯登银行 巴林广场 柏林

竞赛：一等奖，1995
设计：Meinhard v. Gerkan
项目负责人：Volkmar Sievers
合作者：Claudia Abt, Kerstin Dwertmann, Peter Kropp, Brigitte Queck, Werner Schmidt
业主：Schweitzer Grundbesitzund Verwaltungs GmbH & Co. KG
建造时间：1996-1997
建筑面积：11.600m²
建筑体积：59.912m³

由勃兰登堡建造的"前卫客厅"在柏林是具有代表意义的建筑，其位于历史悠久的巴林广场上，现在要打算进行重建，这次重建被视为"批判性的重建"。德勒斯登银行的新办公楼构成了新的封闭式广场的一部分，但是要符合高度限制和设计规范。新楼设计的构思是从总体发展建造起来的。

由浅黄色基调的砂石构成的建筑物立面在垂直方向通过开窗以及青铜色外窗框进行强调。可调节的百叶和发出铜光的窗框嵌在地板上，它们连接起来体现了一种高贵的整体感觉和严谨的气质，并且与基地环境相吻合。

正如什克尔过去的博

物馆中介绍的那样,此设计方案通过圆形与长方形的体块游戏与建筑的历史发生联系。圆形大厅在门厅后面脱颖而出,拥有一个直径31m有玻璃透镜似的屋顶。办公室朝东面向大厅的空地,中间连着一个独立式螺旋形楼梯和两个电梯。与入口层连着的许多阶梯在视觉上降低了大厅的高度,这些阶梯旋转布局形成了一个圆形舞台,从而突出了基本的几何形态。大厅成为一个设计出色、考虑周密,适应各种各样活动的空间。办公区前面设有一条玻璃走廊,赋予了内立面精致的质感。剩余空间被分成一个个小型集会场所,阳光通过走廊空间倾泻于建筑的一层空间。

1 两层高的入口处对称设计与主楼梯和残疾人坡道的不对称设计形成鲜明反差
2 门厅立面细部的照明效果

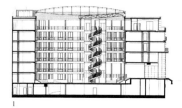

1 剖面图
2 正立面细部剖面图和局部立面图
3 垂直的折叠窗既有遮阳保护作用又活跃了建筑立面

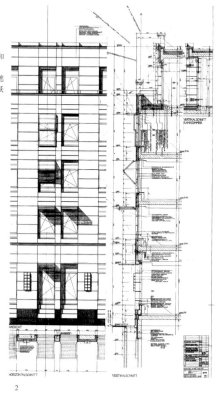

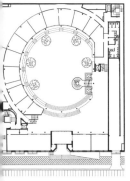

1 连续的阳台铺设玻璃地板，在大厅举办重大活动时可用作展廊
2 屋顶平面图
3 标准层平面图
4 入口层平面图
环状构形成为空间比例的特色
5 入口大厅和接待处
6 会议区
7 人流通道
8 会议桌

5

6

7

8

1 站在被抬高的入口层，圆柱状大厅像一个下沉式露天剧场，营造出接待区和功能厅的辉煌印象
2 透镜状钢结构屋顶，直径有30m，用起重机安装
3 精致的屋顶结构采用可伸展缩钢索组装

1 旋转楼梯采用无框玻璃台阶，通过三个支点支撑
2 每一段楼梯都是悬臂式的，内部和外部的纵梁之间没有构造上的连接
3+4、3/4的房间面向东，朝向圆柱形的"室内花园"并且都有楼梯通向屋顶玻璃墙面

由于外墙三面均为固体结构，日光通过垂直光槽引入室内

1 杰瑟斯·拉斐尔索图设计的"虚像椭圆体"，突出了垂直穿透的空间感觉
2 顶层餐厅采用玻璃隔断
3 会议区像平台一样悬挂在大厅空间之上

2

3

1 上层会议空间的候客区
2 顶层会议室及贵宾接待区内设有钢制壁炉

邦仁再保险公司办公大楼 慕尼黑

竞赛: 1997
设计: Meinhard v. Gerkan
合作者: Johann v. Mansberg, Stephan Rewolle, Eva Wtorczy

这个新建的保险公司总部提供给800名员工使用。他们可以享受多功能设计的灵活布局所带来的愉快的工作环境,同时,委托人要求建筑要显出尊贵的价值体现。工作区的环形设计既能满足功能灵活性又具有独特的建筑意义。各楼层两层一组并且均为4层,这种构形进一步加强了功能灵活性和建筑美感。这一布局也与公司四个部门的机构组织相符合。

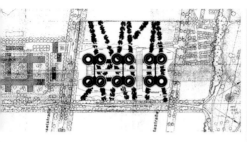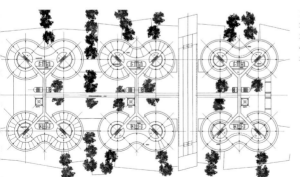

1 组织单元布局平面图
2 总平面图
3 建筑之间的景观设计
4 入口层立面图
5 标准层平面图

电信中心 苏尔

竞赛: 一等奖, 1992
设计: Meinhard v. Gerkan
mit Jens Bothe, Kai Richter,
Hadi Teherani
2.设计: 1993
Meinhard v. Gerkan, Joachim Zais
合作者: Andreas Reich, Stephan
Dürr, Ulf Düsterhöft,
Horst-Werner Warias,
Gabriele Wysocki, Petra Weidmann,
Thomas Böhm

内部空间规划: Wehberg
Eppinger, Schmidtke + Partne
业主: DeTe Immobilien/Koblen
建造时间: 1995-1997
建筑面积: 18.477m²
建筑体积: 59.276m³

由于在概念阶段设计
方案反复更改和简化, 中
标的方案的实现可能会存
在一定困难, 甚至在施工
期间, 设计的顶层平面及

1 入口处设计展现出欢迎的姿态
2 "梳"状布局中突出的建筑体块在施工过程中长度被缩减

1 设备楼梯旁边设有一个备用楼梯
2+3 清晰的建筑形态,建筑物立面采用波纹钢板材料,使建筑独具特色

厅也被取消了。

建筑的主要平面结构形态设计为梳状矩形并且朝南。另有一个环形结构叠加在梳状结构之上,这样它就超出主体屋檐高度一个楼层,显示了它作为建筑群的一个主要部分的显著地位。标准办公区在布局时均朝向中心空间。

两个建筑体块的叠加形成了一个特殊的建筑形态,也缩小了多层人流大厅的体量,这个大厅服务于整个建筑。虽然与整个建筑群相比大厅空间体量有限,但其必须满足流通空间不高于占地面积的20%这样一个前提条件。大厅旁边的客户服务区从主入口进入室内,并未干扰室内相对私密的空间。

建筑的结构体系采用加固混凝土框架结构。地下室墙面材料采用建筑原址的水泥和复原石表层。建筑高层的立面采用大面积波纹金属结构和连续不断的铝合金窗户。

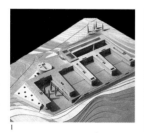

1 竞标设计模型
2 在方案反复变化和体量明显减少之后的建筑设计方案
3 标准层平面图
4 环形建筑结构和线形建筑体块相交叉形成的小型门厅

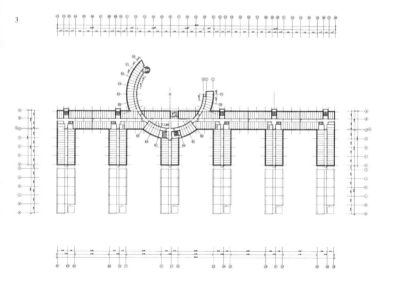

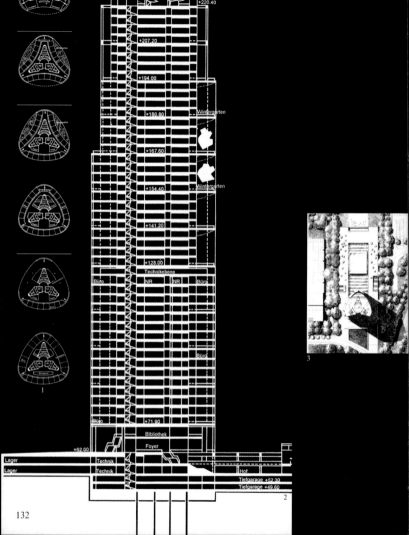

德国邮政通信总部
波恩

竞赛: 1998
设计: Meinhard v. Gerkan mit Sigrid Müller
合作者: Klaus Lenz, Elke Hoffmeister, Bettina Groß

这个建筑的设计与其所处环境融为一体,包括公园内莱茵河畔的一个高层建筑和另外一幢低层建筑。新建的塔楼外形轮廓设计为光滑的曲线,同周边形式鲜明的长条型外观形成鲜明的对比,为标准办公区和综合办公区提供了灵活的布局。楼基应当顺应地势特征并且与公园形成一个有机整体;保留了大量树木。这种新型的平面布局可以允许设计推延到施工开始阶段再完成具体的室内平面布局结构。

1 平面布局
2 剖面图
3 总平面图
4 立面图

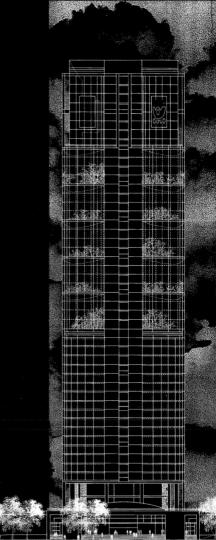

电信中心 霍尔扎赫斯大街 柏林

设计: Meinhard v. Gerkan, Joachim Zais; 1993
合作者: Vera Warneke, Susanne Schröder, Jürgen Stodtko, Gabriele Wysocki, Petra Staack, Stephan Schütz, Doris Schäffler, Dieter Rösinger, Ursula Köper, Stefan Schwappach, Angelika Schneider
业主: DeTe Immobilien
建造时间: 1995-1998
建筑面积: 34.000m²
建筑体积: 128.505m³

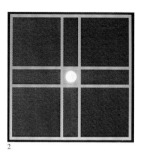

这个建筑是为柏林北部私人设立的分支机构和商业客户服务的新办公部大楼，提供给1000名员工使用。设计方案包括五个相互平行的6层办公楼，建筑的长度紧贴基地边界。这五个建筑由"主楼"连接。室内的特色在于标准办公平面的设计。办公区的点睛之笔在于走廊，它连接着两个最低层的办公空间，主要供客户使用。主楼位于霍尔扎赫斯大街，建筑的设计和尺度限定了入口处的面积，入口面东朝向地铁车站。电信公司员工教育中心也设在此，内院中还设计了一个一层的员工餐厅。

1 入口大厅
2 金属吊顶的构造

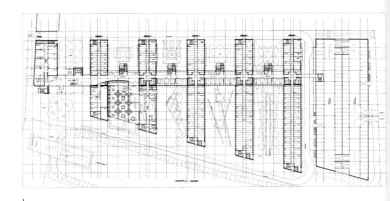

1 一层平面图
2 建筑梳形平面结构形态的鸟瞰图
3 位于霍尔扎赫斯大街的主楼
4 内视深深的庭院

1 梳形结构的连接脊柱采用单面设计。主要流通轴线朝东一面采用玻璃格封闭结构
2 独立式楼梯设计有逃生阳台

1 餐厅与走廊之间的墙面隔断
2 前台
3 餐厅
4 会议室

2

3

4

主楼中走道采用木质护壁

办公区的隔墙装饰有可调节百叶。办公间室内光线可随百叶开闭进行调节

伯特隆技术中心 埃因根

竞赛: 三等奖, 1998
设计: Meinhard v. Gerkan mit Philipp Kamps
合作者: Elke Hoffmeister

一个工业建筑的平面结构形态需要清晰组织流线结构形态,并将其以建筑形式加以体现,从而保证在将来投入使用的高效性。在这个方案中功能有效性在空间组织系统性中得以实现;一个引人注目的前院,经过一个代表性的门厅往前走便是位于大型中心大厅内的工作车间。办公区位于中心大厅两侧,与生产区的联系一目了然。大厅功能灵活,屋顶结构采用构造清晰的组合结构,此外再加上技术与照明设计的完美融合,营造出一个鲜明的整体效果。

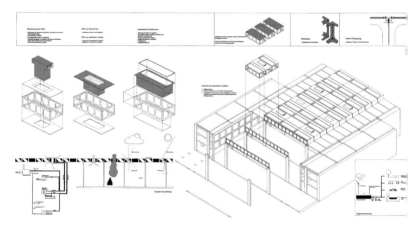

1 总平面图
2 建筑部件和设备配置图
3 从入口处观看整体建筑的模型
4 建筑结构

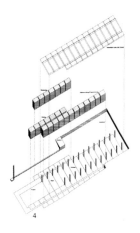

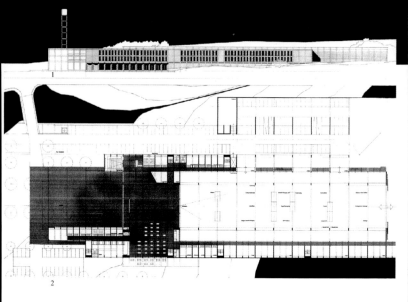

1 纵剖面图
2 一层平面图
3 A-A 剖面图
4 正立面
5 上层平面图
6 B-B 剖面图

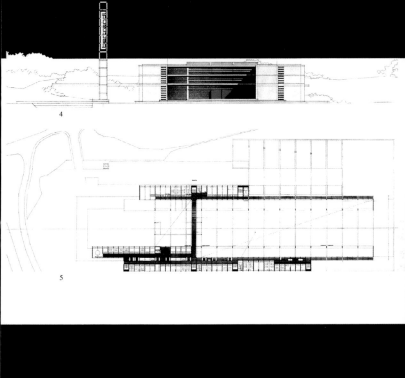

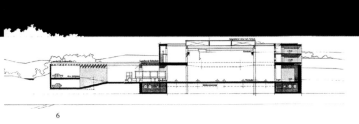

展览馆与消费建筑
EXHIBITION AND CONSUMERISM

1 汉诺威9号展览厅的展示架
2 展厅上层平面图
3 灯泡作为一个设计元素出现，显得极富个性

菲利普博览会展馆

竞赛：一等奖，1998
设计：Meinhard v. Gerkan mit Wolfgang Haux und Magdalene Weiß
合作者：Peter Radomski
业主：Philips Licht, Hamburg
建造时间：Jan.-April 1999
建筑面积：450m²
建筑体积：1.072m³

设计概念

每个展览都是建筑学上令人兴奋的盛会，成为多种多样以及波涛般汹涌的令人兴奋的建筑的展示。与此相反，两层高的"菲利普照明集团"展览馆意在成为一个令人沉思型建筑。外部的封闭引起了公众的好奇并且为内部参观者提供了广阔的空间，使参观者的注意力全部放到了展品上面。展览馆的外形在结构上简洁明快，并且尽可能地减少了材料种类的使用，但在细部处理上却十分精心细致。它独特的外观势必保证菲利普展馆在博览会之后依然让人们记忆犹新。

结构体系

展览馆采用有限部件的组合式体系构造而成。基础体系基于一个 x 轴、y 轴、z 轴为300m大小的网格状空间。主要采用框架，弯曲而坚硬的接点提供必要的承重力，可置换的填充物充当了次要结构，这些填充物根据空间设计可以由开放式或封闭式，透明的或半透明的，发光的或不发光的成分构成。灯泡是作为第一家生产灯泡和当今世界著名的照明设备的先驱——菲利普公司的标志。墙体和室内隔断大多由大量透明的菲利普灯泡组成，灯泡夹置于钢框固定的两片安全玻璃之间。1999年汉诺威展览会展示期间，墙中间一共填充了110000只灯泡。

1

2

1+2 顶层平面,各楼层之间穿透空间
3 主楼层入口

靠近装满灯泡的玻璃墙的
会谈桌
纯钢制成的酒吧
最上层的设计方案
酒吧设计方案和酒吧剖面
图

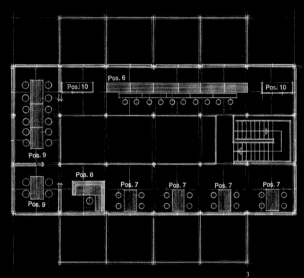

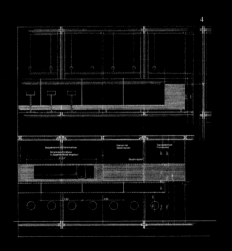

1+2 可移动储物柜带有玻璃抽屉和书册收藏柜
3 主楼层平面图
4 钢框玻璃柜

2

1 可重新组装结构的交叉点
2 原钢材料的前台接待桌
3 "十字形"接点原理

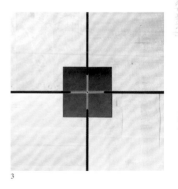

3

里明尼展览中心

竞赛:一等奖,1997
设计:Volkwin Marg
项目负责人:Stephanie Jöbsch
合作者:Hauke Huusmann,
Yasemin Erkan,Thomas Damman,
Wolfgang Schmidt, Regine
Simoneit, Helene van gen
Hassend,
Maria-Chiara Breda, Susanne
Bern, Carsten Plog, Marco Vivori,
Arne Starke

文艺复兴时期的建筑传统,因此设计中体现了线形结构的对称性和清晰的定位感。有四个拱门的门厅(参照古代家族塔的设计)界定出通向展览厅大街的入口。广场两边纵向排列着柱廊和成阶梯状的列柱廊,广场末端是喷泉和有顶棚的门廊。拱形屋顶的廊柱大

1

外联建筑师:Clemens Kusch
静力分析:Favero & Milan,
Venedig,
Schlaich, Bergermann und Partner
景观规划:Studio Land, Mailand
业主:Ente Autonoma Fiera
建造时间:1999-2001
建筑面积:130.134m²

厅成为整个建筑的焦点,展览厅的屋顶采用木结构,由可伸展的钢缆连接,拉紧的木竿承受着水平方向的负重。

作为一个展览中心,设计应该满足功能使用的灵活性,并且要符合高科技标准。基于建筑区位特征,设计应该具有古代和

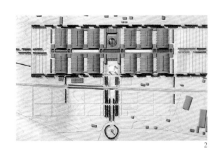

1 入口夜景模型
2 总平面模型
3 菱形屋顶结构展厅的内部空间

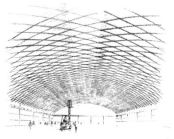

杜塞尔多夫展览馆

竞赛: 一等奖, 1997
设计: Volkwin Marg
合作者: Frederik Jaspert,
Marek Nowak, Olaf Drehsen
Thomas Heuer, Christina Harenberg
Thomas Behr
建造指导: Joachim Rind
合作者: Michael Haase,
Heiko Körner, Marek Nowak,
Simone Ripp, Stefanie Streb,
Petra Tallen, Andreas Wietheger,
Timo Holland, Stefan Menke,
Anne Werrens
静力分析: Schlaich, Bergermann
und Partner; Ing.-Büro Gehlen
景观规划: WES und Partner
业主: Düsseldorfer
Messegesellschaft mbH
建造时间: 1998-2000
建筑面积: 42.500m²
建筑体积: 622.000m³

为了加强展览馆的重要性,设计中在椭圆形的北入口处又增加了两个入口。展览馆成了莱茵河风景的新标志:一个新的入口连接着的欧洲最大的圆形结构的多功能厅,位于莱茵河的轴延伸线上,改建的南入口连接着著名的国会宾馆。

内部空间的分布密度减少了展览馆进一步进行扩建的需要,并且两层高的柱廊搭构出一个室内广场。结构上大胆采用在展览厅中不常见的圆屋顶形态,具有特殊的重要意义。它与威尔海姆克雷斯1926年设计的"托恩哈勒式"展览馆不谋而合,而后者则是第一个博览会的标志性展览馆场。

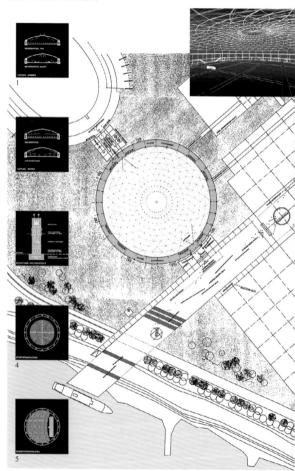

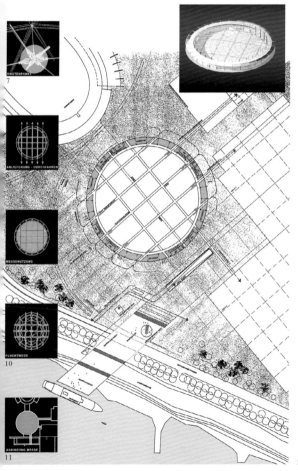

1 夏天的通风装置
2 冬天的通风装置
3 展览馆内的空调装置
4 体育活动场馆
5 音乐会场馆
6 顶层平面图
7 接点
8 运输通道
9 供展览使用
10 安全通道
11 展览馆场地的交通地位
12 一层平面图

"新墙"商业中心4 层 汉堡

专家意见: 1990
设计: Volkwin Marg
项目负责人: Tim Hupe
合作者: Ahmet Alkuru, Frank Hülsmeier, Detlef Papendick, Christoph Berle, Detlef Porsch Christel Timm-Schwarz, Arend Buchholz-Berger, Uli Rösl
业主: Andersen & Co GmbH+C
建造时间: 1994-1997
建筑面积: 7.480m^2
建筑体积: 35.700m^3

首先采用幕墙结构的
建筑是1951—1958年戈
贝尔·尼森设计的一幢
业大楼的四十一层到四一

1+2 "新墙"商业中心的临街立面。这里是汉堡一流时装区
左面：
JIL SANDER 品牌；
右面：
PETRA TEUFEL 品牌。
3 新楼与街区环境
4 旧楼前扩建的新楼
5 新楼与旧楼的遭遇

三层。对这幢建筑的改建和扩建是与前建筑师合作完成的，并且符合市区新增建筑的最新规范，营造出一个极富魅力的现代化购物中心。

透明隔断的购物单元位于室内三个楼层的展廊内，中心是一个玻璃大厅。

波茨坦广场　柏林

竞赛：一等奖，1994
设计：Volkwin Marg
项目负责人：Joachim Rind
合作者：Christina Harenberg, Monika Kaesler, Franz Lensing, Gabriele Mones, Marek Nowak, Efstratios Sianidis, Uta-Eyke Witzel, Robert Stüer, Thomas Behr
业主：GP Fundus Gewerbebau und Projektierung GmbH
建造时间：1996-1997
建筑面积：103.423m²
建筑体积：425.000m³

波茨坦广场没有设计成为通常的缺少商业气氛的内向型，而是营造了一处风格简洁的购物中心。购物中心上覆盖着长长的玻璃顶棚，从而使这里成为庆祝活动、美食以及在旁侧店铺购物的核心地带，同时这里也成为通向里面专业市场和专卖店的入口通道。它在临巴赫夫大街立面呈现出自由写意的设计风格，而在沃赫水域的新建的住宅区则拦断了主河流的流动。建筑物的高度与周围的建筑保持一致，从而与周围环境的平面结构形态相符合。住宅单元与建筑整体保持统一，内部结构简洁。在临水建筑带的西侧新建了一座3层高开放式布局的建筑，充分体现了区域风格。

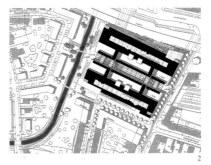

1 拱廊购物中心的一侧
2 总平面图
3 主入口

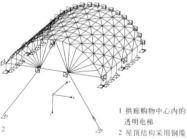

1 拱廊购物中心内的透明电梯
2 屋顶结构采用钢缆支撑
3 精确的构造强调出屋顶结构桶状拱顶特征

交通设施
TRANSPORT

1 新火车站鸟瞰图
2 横剖面图
3 站台大厅夜景

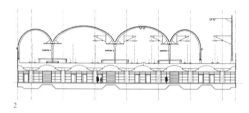

柏林—施潘道 火车站

竞赛：三等奖 1993
设计：Meinhard v. Gerkan
合伙人：Hubert Nienhoff
项目负责人：
Sybille Zittlau-Kroos,
Birgit Keul-Ricke, Elisabeth Menne
合作者：Almut Schlüter,
Peter Bozic, Agnieska Preibisz,
Margret Böthig, Diane Berve,
Karl Baumgarten, Matthias Wiegelmann,
Gerd Meyer, Andreas Dierkes,
Paul Wolff, Marietta Rothe,
Kerstin Struckmeyer, Peter Schuch,
静力分析：
Schlaich, Bergermann und Partner
业主：Deutsche Bahn AG
建造时间：1996-1998

车站大厅面积：4.891m^2
屋顶面积：20.607m^2
建筑体积：28.369m^3

施潘道火车站是汉诺威—柏林主干线公共设施的一部分，它和施潘道的市政大厅一样地位突出。建筑的一层拱廊设计成为一个具有商业功能并且提供火车站相关服务设施的空间。主厅长430m，完全由平行的拱形屋顶覆盖，这种设计形态引起了人们对传统车站大厅的联想。就好像展览馆里展示的1996年在威尼斯由德国巴

3

1

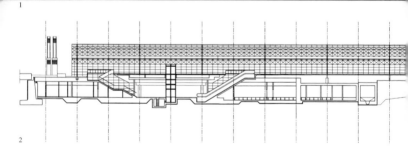

2

尔公司设计的"文艺复兴时期的火车站"。虽然设计形式召唤传统，但是构造科技还是要体现21世纪风格。屋顶结构由一条纵向的梁体支撑，梁体是由沿站台中轴线间隔18m分布的柱体支撑。每根支柱的横轴都辅以曲状加固拱肋。因为站台复杂的几何构造，因此玻璃结构的外墙有着轻微的变化，但整体看上去仍然是一样的。整个结构体系通过拉缆加固，拉缆斜向加设于屋顶下方。间隔3m的悬臂吊灯在夜间为站台提供照明，在玻璃拱形圆顶和向上的梁体的反光给人一个印象深刻的美妙的钢与玻璃结构的空间感觉。

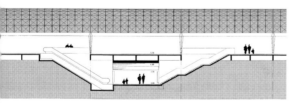

1 玻璃大厅长达430m,是现今欧洲最长的玻璃大厅
2 车站大厅剖面图
3 加伦街地下通道的剖面图

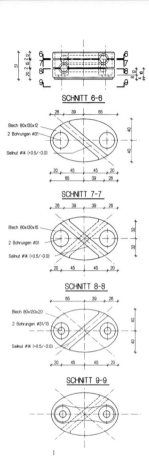

1 由拉缆结构钳紧固定部件细部
2 站台建筑的设计也采用轻钢玻璃结构
为了调节大厅内光线氛围，人工照明精心设计达到了玻璃拱形屋顶的多种反光效果

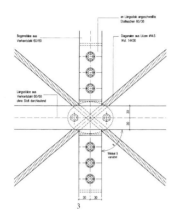

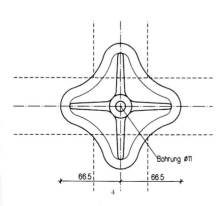

1 屋顶内侧细部。所有的节点都是灵活的。因此表面上的矩形格子能够随着弯曲的几何构形进行调整，造成半径不一的弯曲形状。没有任何网格构形和相应的玻璃结构是相同的
2 屋顶由12700个不同的窗玻璃片组成
3 节点的内部构造
4 节点上盖钢板，由中央螺钉固定

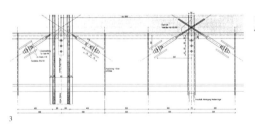

1-3 斜向拉缆终端锁扣固定
4 玻璃屋顶的最大拉伸弧度

1+2 站台下面的车站大厅的顶棚是正弦曲线构形。间接光源设计弥补了室内有限的高度
3 通向站台层的楼梯
4 车站和站台大厅的剖面图

火车站，什未林

专家意见: 1998
设计: Meinhard v. Gerkan
合作者: Arend Buchholz-Berger, Oliver Christ, Justus Klement, Michael Biwer

1+2 建筑采用箱形外观，它的外观特色与火车的动感形成了对照
3 车站中央大厅内视图

火车站位于城市郊区，一排排刚种植的呈环状布局的树木勾勒出基地的边界。

车站建筑本身位于铁轨之上并且与抬高的高速列车轨道融为一体。与蒸汽高速列车相比，车站像一个水晶盒子，由一个长跨度的混凝土结构体系支撑，镶嵌以玻璃结构，这与高速列车的流线型外观形成强烈反差。一个两层的大厅位于纵轴线中心区域，充分挑战现代交通设施科技，呈现出透明、开阔的空间以及清晰的人流流通组织。

1 车站南面入口
2 车站位于什末林郊区的空阔环境中,环形布局的绿化界定出车站区域的边界 建筑的结构框架内镶嵌着玻璃砖,有利于光线的通透
3 车站入口与一层平面
4 站台层13m标高

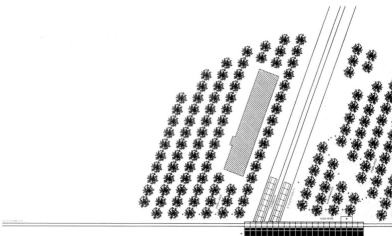

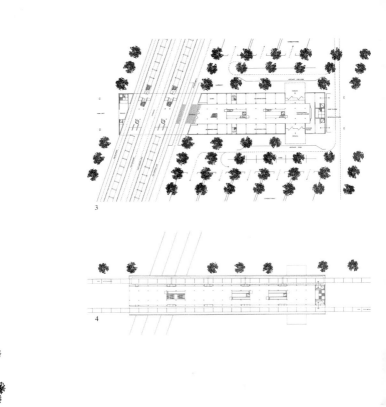

勒尔特火车站 柏林

竞赛：一等奖，1993
设计：Meinhard v. Gerkan
项目合伙人：Jürgen Hillmer
项目负责人：Susanne Winter, Klaus Hoyer, Arend Buchholz-Berger, Hans-Joachim Glahn
合作者：Prisca Bucher, Christel Timm-Schwarz, Gisbert v. Stülpnagel, Andreas Ebner, Bettina Kreuzheck, Vita Römer, Elisabeth Mittelsdorf, Ralph Preuß, Sabine Gressel, Stefan Bachmann, Constantin Dumas, Michael Scholz, Werner Schmidt, Hubertus Pieper, Peter Karn, Thomas Weiser, Gisela Koch, Sebastian Geiger, Peter Krüger, Christian Kreusler, Lothar Scharpe, Monica Sallowsky, Amra Sternberger, Radmilla Blagovcanin, Hans Münchhalfen, Antje Lucks, Maike Carlsen, Claudia Gern, Saban Yazici, Tomas Nowack, Markus Siegel, Diana Kurscheid, Wolfgang Höhl
静力分析：Schlaich, Bergermann und Partner; IVZ/Emsch+Berger
照明设计：Peter Andres +Conceptlicht GmbH
建筑技术：Ingenieurgesellschaft Höpfner
业主：Deutsche Bahn AG vertreten durch die DB Projekt GmbH-Knoten Berlin
建造时间：1996-2005
建筑面积：180.000m²

在具有历史意义的"勒尔特——巴赫霍夫"即将建成一个德国最重要的火车站，这里将实现内城高速列车的东西向和南北向连接。火车站南北向位于斯普雷河与蒂尔加藤河地下15m，东西向被抬升，它的长达430m的站台的屋顶采用轻钢玻璃结构，两个条块状建筑从站台大厅穿插而过，清晰地界定出建筑在城市环境中的南北向轮廓。中央大厅宽50m，长170m，位于这两个建筑条块之间，同样采用精致的拱顶结构。大厅成为位于一侧的莫尔斯区进入车站的入口，同时也是位于另一侧行政区进入车站的入口。这两个条块状建筑高45m，连接车站大厅横跨一处70m宽的空间，预计用于服务区、办公区及车站旅馆。火车站的玻璃屋顶采用拉缆结构体系，组合成1.2m×1.2m的模块。拱顶、梁体和斜向拉缆的结合构成了建筑的屋顶结构。建筑侧面的加固结构采用间隔20~30m的刚架。

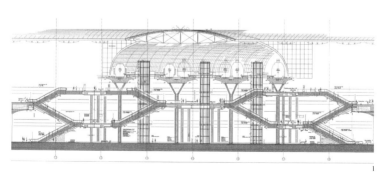

1 车站大厅纵剖面图
2 车站南面入口

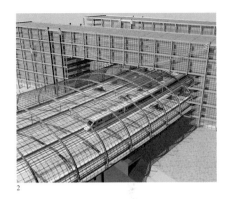

1 施工现场鸟瞰图，1999年5月
2 两个条块状建筑成为车站大厅的东西向连接
3 车站大厅内视图
4 拱顶结构的东西向车站大厅内视图

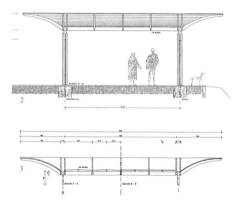

5

1 结构体系组成部分
2 立面图
3 剖面图
4 汉诺威——奥特兹站台
5 屋顶边缘设计细部

S-巴赫车站 汉诺威

设计：Meinhard v. Gerkan,
Jürgen Hillmer; 1996
合作者：Klaus Nolting
业主：Deutsche Bahn AG
建造时间：1997-1998, derzeit gibt es 7 Dächer
场地平面：9 × 9m

S-巴赫车站可被用于不同的"电影剧本"，这就是建筑师在设计中为什么会选择一个没有明确方向定位的多入口的基本正方形结构。站台区由一个长宽各为9m的平屋顶所覆盖，由四根破旧的柱体支撑，支柱呈十字形布局，相隔间距为5.10m。屋顶结构采用瓦楞金属板，由德国巴赫公司专制。这个屋顶结构呈现出轻微的弧度，成为了车站的"活泼的衣领"。站台地区的照明设计主要采用地灯，安设在支撑屋顶结构的支柱之间。站台边缘选用浅色材料铺地，与呈上翻弧度的屋顶结构内侧的金属板相互反射，相映成趣，在夜间勾勒出站台轻快上扬的体态。

泰尼勒夫机场

竞赛: 1997
设计: Meinhard v. Gerkan
合作者: Klaus Lenz,
Nicolai Pix, Maren Lucht

这个设计整体构形统一,赋予泰尼勒夫机场别具一格的特色。航站楼主楼的凸形结构与侧楼的凹形结构的体形游戏加强了中央建筑的重要性并且优化处理了航站楼的停车区域。曲线和直线的构形处理为到达或出发的旅客营造出宽敞的空间感和清晰的方位感。

大厅的屋顶结构的设计成为飞机在飞行时产生的失重感的象征性表达。屋顶的"帆状结构"在建筑设计和构造特征上都可谓是独树一帜,它向外伸展像幕布一般遮饰于中央区域,给人一种轻快溢满的膨胀感。这个结构采用纵向拱顶和光滑的横剖面表面材料。

航站附楼模型

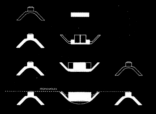

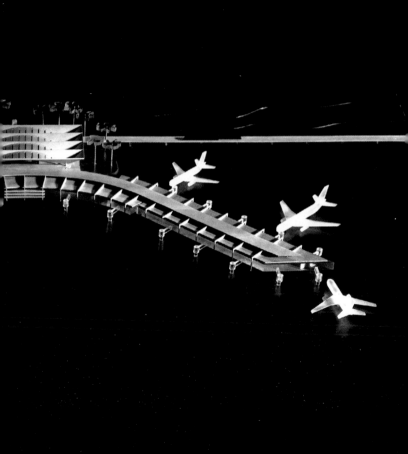

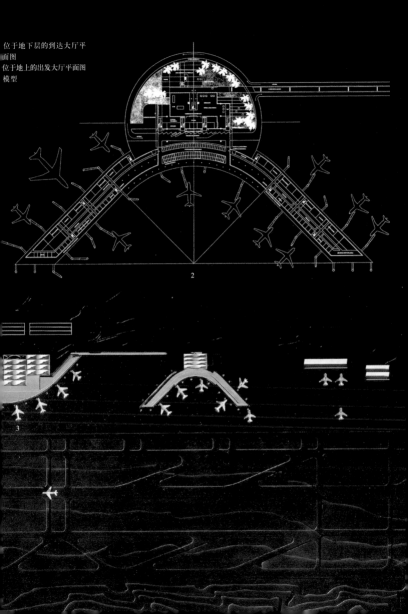

位于地下层的到达大厅平面图
位于地上的出发大厅平面图
模型

慕尼黑空港 2 号航站楼

竞赛: 1998
设计: Meinhard v. Gerkan mit Klaus Lenz
合作者: Nicolai Pix, Sigrid Müller, Ralf Schmitz, Bettina Groß, Philipp Kamps, Eva Wtorczyk, Anne-Kathrin Rose

1

从形式上来说,原来的飞机场是一个独立自主的综合体,因而扩建时不能采用非传统的建筑形式。基于这一点,再加上在设计中对功能灵活性的一贯要求,建议扩建部分采用组合体系。这些组合模块可以在设计中忽略不计也可以加以使用或相互置换。这种结构设计原理允许空间的灵活扩张或改换使用功能。这种组合结构体系只是提供了一个结构框架,只对主要功能空间加以铺设。建筑的最终完成包括立面设计和屋顶结构在内都应与实际需求保持一致。在这个统一的单元构造体系中,形式没有决定功能,而是功能在结构框架中丰富自己。

3

1+2 模型显示出设计概念中的组合构造体系
3 组合结构体系体现出乘客通道在空间上独特的外形特征

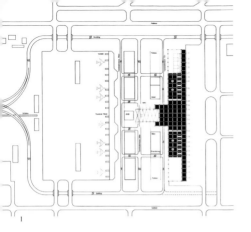

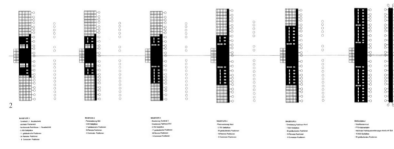

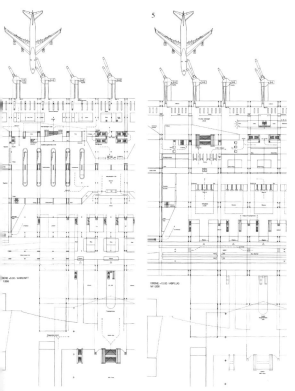

1 总平面图
2 施工工期
3 出发大厅的内视图
4 位于地下层的到达大厅平面图
5 位于地上的出发大厅平面图
6 横剖面图，包括邻近的慕尼黑机场中心的屋顶结构

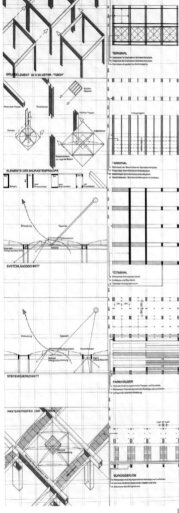

1 组合结构体系图解
2+3 屋顶支撑结构的模型研究

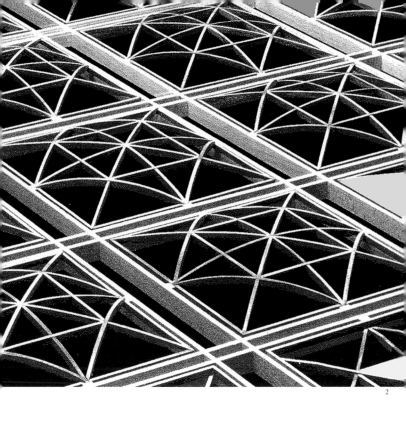

2

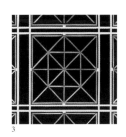

3

斯图加特空港3号航站楼

竞赛：一等奖，1998
设计：Meinhard v. Gerkan
合作者：Klaus Lenz, Sigrid Müller, Stephan Rewolle, Oliver Christ, Hito Ueda, Volkmar Sievers, Otto Dorn
业主：Flughafen Stuttgart GmbH
建造时间：2000-2004
建筑面积：53.300m²
建筑体积：384.300m³

1+2 扩建之后，屋顶部分树形结构柱支撑，而且虑到建筑的高度限制，会逐渐降低。
3 1号航站楼到3号航站楼顶结构的连接
4 横剖面图
5 屋顶构造

树形布局的支柱和堤状建筑剖面形态都成为设计中富有特色的元素，在扩建中将予以保留。基于空中安全考虑，建筑高度有所限制，因此不能沿用现有的屋顶高度。单柱支撑的"伞形屋顶"在高度上逐渐降低，符合了这一先决条件。对增加建筑深度的要求也通过主楼前的平板得到实现。在临街的那一面，260m长的入口雨篷与屋顶主结构相连。

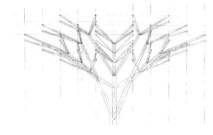

1

3

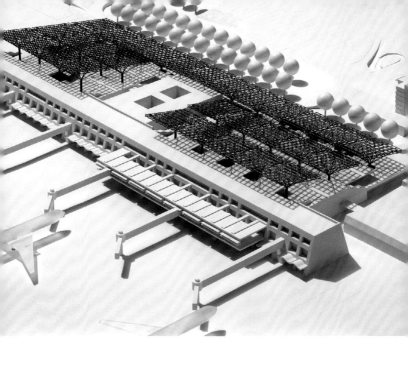

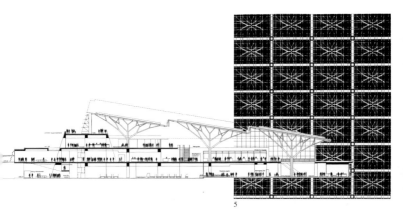

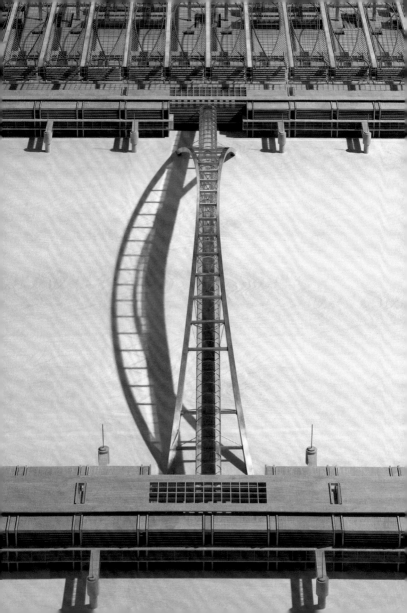

勃兰登堡国际机场
柏林

竞赛：一等奖，1998
设计: Meinhard v. Gerkan,
Hubert Nienhoff, Joachim Zais mit
Hans-Joachim Paap
合作者: Alexander Buchhofer,
Carsten Borucki, Christian Eling,
Stefan Friedrichs, Marc Gatzweiler,
Anne Harnischfeger, Sven-Eric Korff,
Jörn Orthmann, Nicolas Pomränke,
Cornelia Thamm, Olaf Timm,
Monika van Vught, Gabriele Wisocky

2

设计概念基于这样一个认识之上，即飞机场必须被看作是一个"过程"，而非固定的状态。因此有必要建造一个富有活力的框架为以后的扩建留有余地。因而设计在赋予个人充分的自由思考空间的同时，为以后的建设提供了一个总策略。增长和变化成为其充分条件的首要解决方案，两者同等重要。从设计概念上说，为德国首都设计的新的机场大厅，作为新机场设施中的主要"脊梁"元素将成为东西向平行排列的出发大厅和到达大厅的中轴线。这就限定了主要的人行流动组织。

到达候机室与出发候机室设在航站楼中心大厅，另外还有一个备用"离埠"通道。一个乘客天桥提供了到机场巴士的连接，作为一个不同寻常的结构，这个过道天桥是新机场最有特色的部分。

乘扶梯一到达出发大厅，户外的机场跑道就可一览无疑。设计采用组合模块体系，可根据需求添加。航站主楼分为两层，因

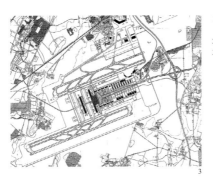

1 人行天桥连接着航站主楼和附楼
2 整体内部工程的模型
3 整体内部工程的平面图

3

1

此在地面一侧形成了出发大厅和到达大厅地上地下的分隔。出发区被建成一个有充足光线的大厅,并被细工精饰的轻钢玻璃屋顶结构覆盖。在位于地面一边设计了一个多层停车场,将通过带有遮蓬的人行天桥和电梯与机场相连接。停机坪前最先映入眼帘的是一片绿地。一期工程的停车场设计标高与街道标高相同,紧靠航站大楼,并与一个往返巴士线连接。

2

- 悬臂式雨篷下的下客区
- 人行天桥连接停机坪和跑道，跨度长达300m
- 能看到停机坪的出发大厅

3

1

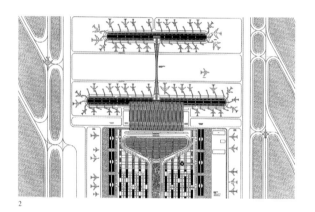

2

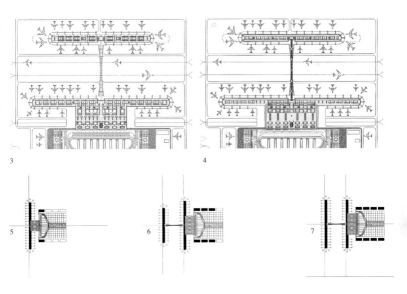

1+2 航站大楼二期工程完成。
下客区位于多层停车场一侧
3 到达大厅位于地下层
4 出发大厅位于地上一层
5-7 各期工程完成的情况

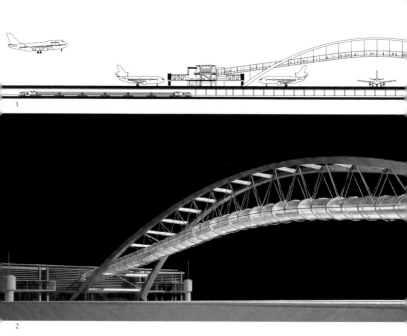

1 纵剖面图
2+3 人行天桥好像玻璃管悬挂在一个拱形结构中
4 乘坐电梯到航站附楼的长长的距离，可以通过在人行天桥上看到的景观得到调节。这个设计同标准连接通道相比堪称世界第一

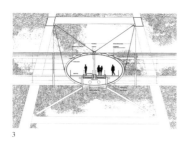

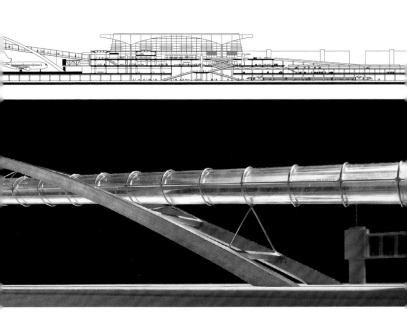

4

215

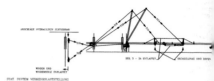

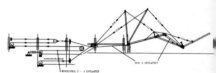

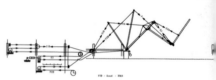

折叠式桥 基尔—霍恩

专家意见: 1994
设计: Volkwin Marg mit Jörg Schlaich
合作者: Reiner Schröder, Hito Ueda, Anne-Kathrin Rose, Dirk Vollrath in ARGE mit Schlaich, Bergermann und Partner, Jan Knippers
业主: Magistrat der Stadt Kiel
建造时间: 1996-1997
长度: 116m, davon ca. 25 m als Faltbrücke gebildet
宽度: 6m

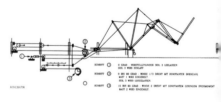

霍恩河东岸的原修船厂一带直接面对城市中心和火车中心站。这个地区作为"千禧年工程"的一部分将被重建成为一个集住宅、办公及商业为一体的新区。作为新区的一部分，戈登区

216

1 桥梁叠起的状态
2 采用机械手段驱动桥梁折叠过程的各个阶段
3-6 多彩用色使得各部分的结构视觉化

4　　　　5　　　　6

将通过在霍恩河上建一座更具灵活性的新桥与市中心连接。双折活动桁架桥或者折叠桥是造船工业中一项通用技术特点,这个创新发展的三折活动桁架的桥给市民们创造了一个鼓舞人心的海上象征。建筑学、工程学以及动力美学同时有效地融入到了一个新的令人振奋的标志中。

5

哈韦尔河跨河铁路桥
柏林—施潘道

设计: Jörg Schlaich mit Meinhard v. Gerkan; 1994
施工: Schlaich, Bergermann und Partner
建造规划: Arbeitsgemeinschaft Havelbrücke Spandau
上部结构建造: Krupp Stahlbau Berlin GmbH

业主: Deutsche Bahn AG
建造时间: 1996-1997
长度: 124.2m
宽度: 47.5m

随着东西部联系的发展，尤其是铁道的扩建，连接了汉诺威与柏林，实现了市际高速列车，建成了七条跨哈韦尔河的铁路。

如果德国铁路桥的构思是采用一个标准化的大型拱桥式的结构，那么将会是十分的不相称，而且还会严重影响连接施潘道城市的环境。与之相反，工程学与建筑学的人员进行合作，深入研究，创作出一个堪称经典之作的悬桥，与桥梁两端相连的拉索结构成了桥梁的基本形态。这座桥槽如波纹般向上突起，考虑到静态力量的运行状况，三个桥槽上的波状突起面则决定其剖面高度。

1 六条平行梁使得大桥可以分期施工
2 建造原则与拱桥相反。顶弦是拉紧的，所以最高点高于支撑点，而且不是位于中心点，这样从桥上看河的视野也就不被阻挡了

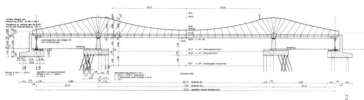

1 图片前景部分：铁路的路堤带有防噪声设施
2 构造剖面
3 桥梁拱肋的构造分隔清晰可见。建筑各部分的排列犹如肋骨那样精细

沃布利茨跨河大桥 波茨坦

竞赛: 1997
设计: Meinhard v. Gerkan mit Gregor Hoheisel

在 2001 年的联邦公园展览中,波茨坦要与"波茨坦森林和哈弗尔河流域的种植景观区"连接起来。作为第一个标志性建筑,这座桥横跨沃布利茨河,将波茨坦和特普利茨岛相连接。

桥的主体结构和扶手采用连续的格子横梁排列,而且用木板条覆盖起来。另外水中还立起木杆,用作路标,道路两旁的树也使得建筑结构得以延续。

1+2 桥上的发光立柱如同大道两旁的路灯
3 桥梁剖、立面图

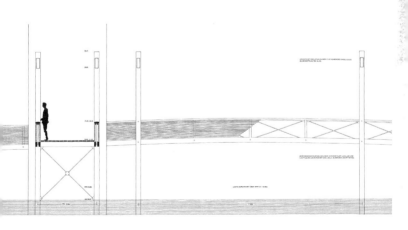

快速列车的内部设计

竞赛: 1997
设计: Meinhard v. Gerkan, Jürgen Hillmer
合作者: Renata Dipper, Stefan Bachmann, Thilo Jacobsen

延续,而在视野上则强调为窗户的纵向排列。线性的座位和走廊使这一设计原则成为可能。在细节上,使用的材料有玻璃、金属、木材和皮革,使得气氛既

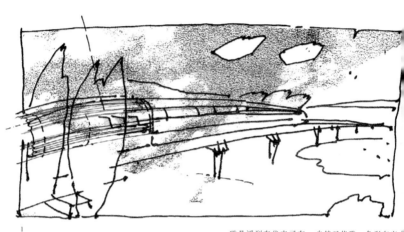

磁悬浮列车代表了有轨交通的新技术,因而必须用一种简洁的设计概念来表现,让社会各个层次的人都能理解。磁悬浮列车在进步、高速以及舒适上都有一定的表现,但设计原则的精髓可以总结为一句话:"总体整齐划一、细处风格迥异"。高速运输被看作是一个可持续的、统一的实体。列车外壳上的动态形象在内部得到了

自然又优雅。色彩在室内极少使用,除非是用作区分标志。自然光和人造光的妙用,相得益彰更显设计之永恒性。

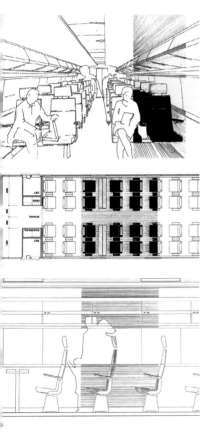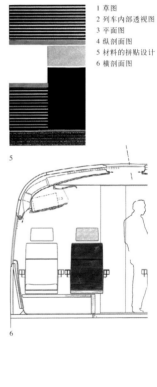

1 草图
2 列车内部透视图
3 平面图
4 纵剖面图
5 材料的拼贴设计
6 横剖面图

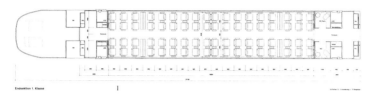

Endsektion 1. Klasse   1

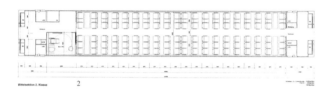

Mittelsektion 2. Klasse   2

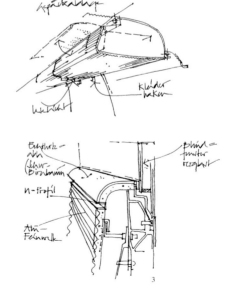

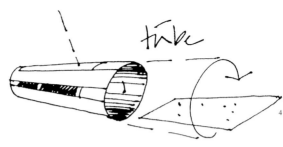

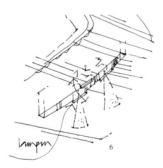

1+2 平面图
3 细部草图
4+6 概念草图
5 车厢序列

"大都会"快速列车室内布置

竞赛：一等奖，1996
设计：Meinhard v. Gerkan, Jürgen Hillmer
项目负责人：Renata Dipper, Birgit Föllmer
合作者：Susan Krause, Frank Hülsmeyer, Maja Gorges, Kristina Kaib, Bernd Stehle, Torsten Neeland
照明设计：Conceptlicht GmbH Helmut Angerer
业主：Deutsche Bahn AG
建造时间：1998-1999

"大都会"快速列车是一种新型的列车交通系统，提供两个城市间的往返连接，其高标准的舒适程度和服务旨在分流飞机的一部分业务。所以其设计集中在创造列车的特殊形象，外部设计同内部设计一样，都要被看作一个紧密的单位。

这个设计从"市际"高速铁路等同类产品中凸现出来，形成了一个独立的"列车形象"。"大都会"快速列车就是"银质列车"，外壳的金属特性代表了速

度和质量。其动态形象在内部得到了延续,那就是水平剖面和连续成排的窗户。线性结构的座位和走廊区域增强了这一印象。车厢之间又依据"办公区"、"俱乐部"、"安静区"划分开来,给乘客提供了各种不同的功能用途。所有的车厢在一边设有两排座位,在走廊另一边则是一排座位,所有的座位都采用一个样式,集成折叠桌,并提供通讯连接。这些装置的一个重要特点就是天然材料的应用:多层夹板、不锈钢以及皮革营造了一个舒适的氛围;这些材料的使用也随着时间的推移而更显光彩。

在材料上尽量避免了塑料的使用,特别是着色概念上的简化,只表现出材料间的天然纹理差别。行李放在窗户上方的架子上,架子旁边集成设计了间接照明灯暨阅读用灯。

2

3

4

前页:
乘客车厢内视。顶棚材料选用网状金属板,它的另一个用途就是充当整体照明概念中的反射装置

多层座椅设计,座椅上有瑞士花梨木饰面,还覆盖着黑色皮垫

-4 座椅侧板选用螺纹钢材和瑞士花梨木制成。不锈钢行李架与整体照明装置设计在一起

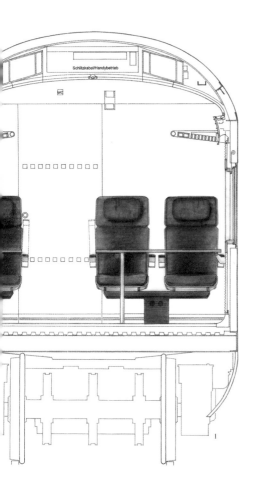

1 车厢横剖面图
2-3 四个座位中间安设一张长折叠桌
4 垂直方向可调的皮质头垫

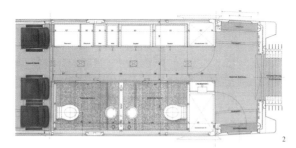

1 车厢入口。楼梯上的间接光源。不锈钢扶手。花梨木墙板上镶有不锈钢条作为保护。这个墙面还有一个作用,就是储放物品。地板上铺有羊毛地毯
2 局部平面图,带有厕所、储物区以及报架
3 厕所内部:不锈钢、石材以及镜子
4 厕所入口和报架间的过道

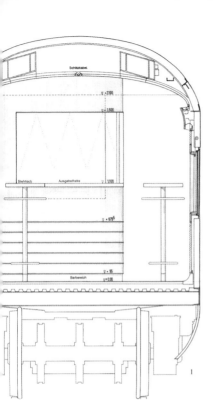

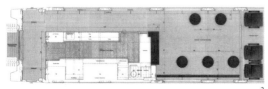

1 餐车剖面图
2 餐车平面图
3 餐车内部,设有吧台。餐台旁边是镀铬钢制百叶窗

政府和社区建筑
STATE AND COMMUNITY

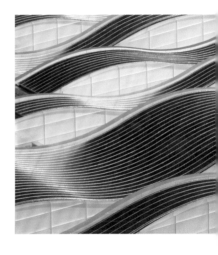

1+2 模型及草图标明的总平面
3 巴丁广场及胡志明墓到阵亡战士纪念碑有一个轴向上的连接，这个连接由左边的国会建筑和右边的议会建筑作为框架构筑而成

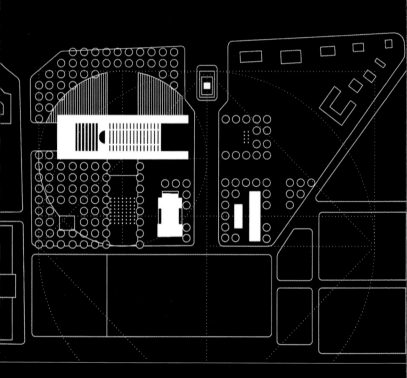

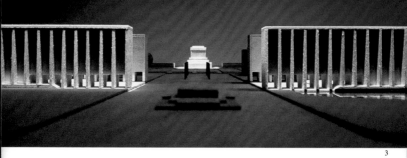

3

议会大厦　河内　越南

竞赛: 1998
设计: Meinhard v. Gerkan
mit Nikolaus Goetze
合作者: Oliver Christ

一期工程中，国会大厅将被特别建造在胡志明墓到阵亡战士纪念碑走向的主轴上。二期工程中，随着新议会大厦将多出一条横轴。两部分互相关联，由纵轴从中间截开，组成一个巨大而又简单的图案。在两个相对的建筑之间形成了一个开放的"议会广场"，所以从这两座建筑都可以进入。有一个变化就是议会大厦坐落在国会中心的南部。

1

3

2

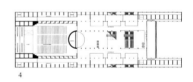
4

1 纵剖面图
2 环形部分二层大厅平面设计图
3 南立面图
4 入口层大厅平面图
5 纵剖面图
6 环形部分议会大厅层平面图
7 南立面图
8 入口层议会大厅设计图

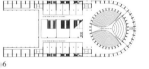

另一个设计方案建议将建筑两个部分均建在西面
1 西南立面图
2 一层平面图
3 包括胡志明墓在内的总平面图

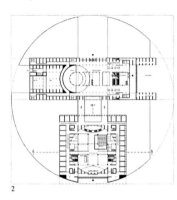

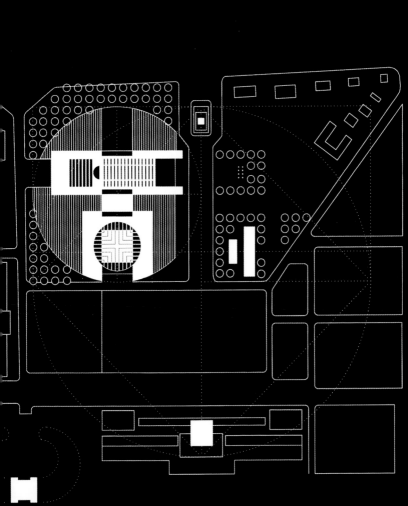

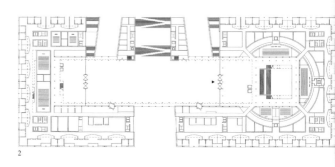

1
2

法院 安特卫普 比利时

竞赛: 1998
设计: Meinhard v. Gerkan, Klaus Staratzke
合作者: Oliver Christ, Pinar Gönul-Cinar

这座建筑必须面对的矛盾从本质上说就是：在城市边缘、高速公路的路口要建一座具有代表性特点的庭院式建筑。为了克服这个困难，这座建筑对指向城市的高速公路入口做了新的诠释，它穿过路口，与道路成直角，如同"城市大门"。从建筑物的入口可进入内城区，从这里还可以进入两座分开的不同功能建筑部分。主楼向大众展示的外貌，如同一座玻璃塔。

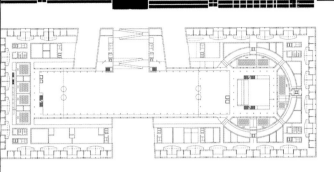

南立面　　　　　4 标准层平面
顶层平面　　　　5 模型参考
纵向剖面

市政厅 布郎什

竞赛: 二等奖, 1997
设计: Meinhard v. Gerkan mit Philipp Kamps

这是一座历史性建筑, 曾被用作单身工人宿舍, 是布朗什新市政厅的"心脏"。它以前的特征仍被采用, 并在精心"修饰"后融入了新的设计中。虽然体量上增大了, 但扩建部分的设计与旧的建筑以某种方式紧密连接, 珠联璧合, 似乎表现出对历史的尊敬。与此同时其建筑设计又见证了过去与现在的统一。虽没有刻意的历史化, 但是仅仅是旧与新的毗邻就衍生出了一个新的实体, 具有其自身的特性。

建筑坐落于哈瑟特斯, 主楼梯就是入口的中心地带, 市政厅里任何么

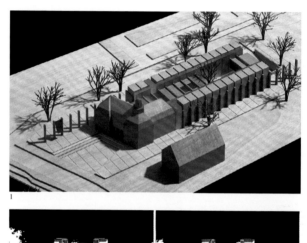

1

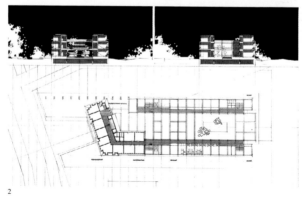

2

众区域都可以由此直接进入,后面紧挨着的小区域就是市民中心,有着玻璃屋顶。历史性建筑保留了其地域和传统上的特点,而新建筑又为市政厅创造了条件,起到社会中心的功能。

1 模型
2 剖面图和上层平面图
3 总平面图
4 正立面设计细部
5 纵剖面图和一层平面图

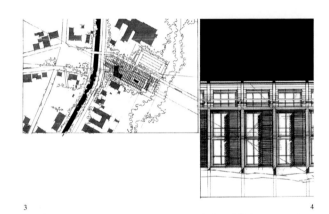

3

4

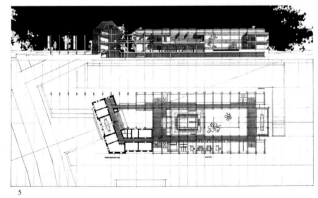

5

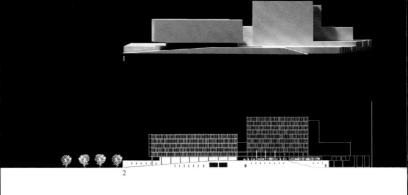

警察总署　丹斯特杜伊斯堡

竞赛: 1998
设计: Meinhard v. Gerkan
合作者: Michael Biwer, Nicolai Pix

杜伊斯堡林区原先是一处狭窄而又有限的地方，设计建议建造标准办公区和一个综合性停车服务中心。停车服务中心位于底层，它的顶上就是办公区。为了与木材堆叠和浮筏结构进行类比，扩建部分可进行横向的水平分隔以及垂直向的剖面分隔。堆叠和分层设计原则给一个建筑项目引进了内外有别的设计内涵，而其本身功能却毫无变化。

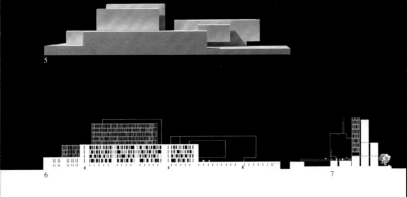

1 模型南立面
2 南立面图
3 用树干的浮筏结构设计，与林区区域特色相吻合
4 总平面图
5 模型北立面
6 北立面图
7 西立面图
8 标准层平面图
9 正立面细部剖、立面图

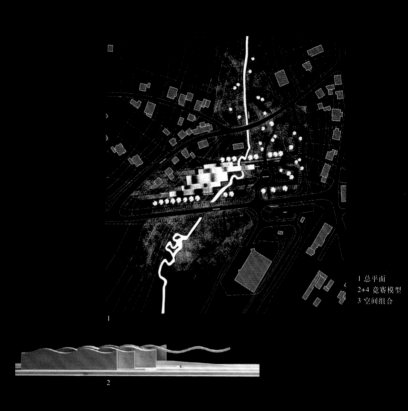

1 总平面
2+4 竞赛模型
3 空间组合

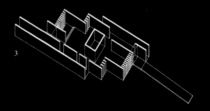

"坏小子" 史特本游乐场

竞赛: 一等奖, 1997
设计: Meinhard v. Gerkan mit Anja Knobloch-Meding
合作者: Sigrid Müller, Sona Kazemi
建造指导: Anja Knobloch-Meding, Justus Klement
合作者: Bettina Groß, Evgenia Werner, Maja Gorges, Jessica Weber, Marina Hoffmann
业主: Gemeinde Bad Steben
建造时间: 1999-2001
建筑面积: 4.216m²
建筑体积: 20.038m³

从汽车道和停车场出来有一条缓缓向上的斜坡，穿过史特本水渠通往一个入口处，入口的上方是一个长长的抽出式屋顶棚——给人一种欢庆节日的喜庆氛围。当然也是一种特殊场景的标志，这在赌场来说是很明显的。这种内涵在交错带状的波浪状屋顶中也得到了体现——其象征意义就是在赌博时运气好坏参半，但不代表对赌博(潜在的)沉迷。波浪状屋顶结构从视觉上来说减小了建筑的体积，这在游乐胜地是很少见的。

一楼的构成与屋顶的带状结构很相称，有成排的平行空间，每一排都有各自的用途。非公众空间在二楼，技术部门设在地下室。

小镇里的赌场——有趣而且有一点奢华，但是严格而又有序，与游戏规则相应。

1 入口处
2 侧立面图
3 一层平面图
4 上层平面图
5 纵剖面图
6 横剖面图

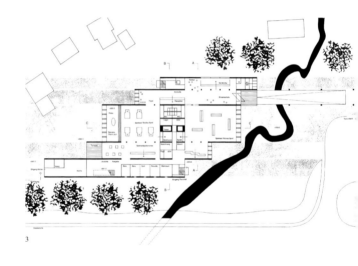

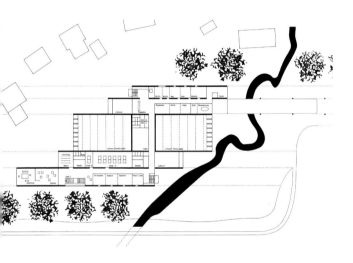

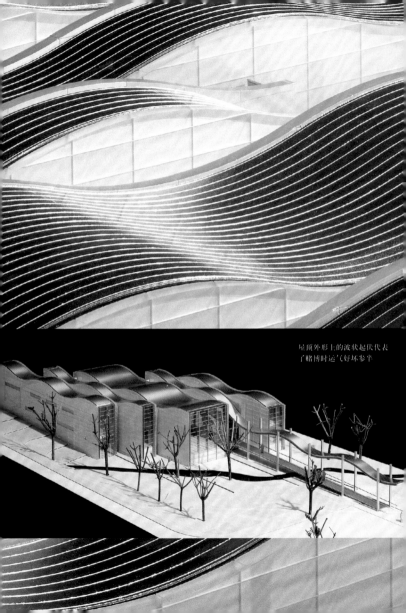

屋顶外形上的波状起伏代表了赌博时运气好坏参半

健康中心　慕尼黑

竞赛: 1997
设计: Meinhard v. Gerkan
合作者: Klaus Lenz, Johann v. Mansberg

新建筑有三面采用了周围现有建筑的方块式结构,融入了整个街景之中,但在靠里的一面却空出了一片宁静的户外空地。楼体在结构上清楚地分为三个护理单位,其中病房是垂直排列的。这种效果避免了太长的医院走廊,鼓励病人倾向放松。一致的方块式建筑反映出了公共空间的布局,其间的露台又可以用来为下面的楼层提供照明。

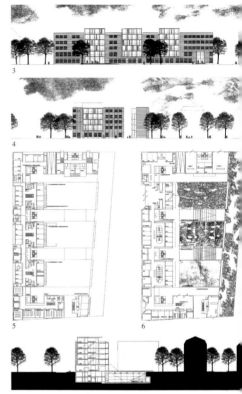

1 模型
2 总平面图
3 西立面图
4 南立面图
5 上层平面图
6 一层平面图
7 横剖面图

爱心医院 德累斯顿大学

竞赛: 1997
设计: Meinhard v. Gerkan
合作者: Johann v. Mansberg

新方案中内院的大橡树组成了建筑群的一部分，并提供了一个中心点。在其南面，入口大厅朝着橡树的方向打开，同时在通向道路的一面，这个大厅又是主要的公众入口。从结构上和建筑学上来说，门诊部同检查室和治疗室一起分别设立在底下两层楼中，看护病房安置在上面，即三楼。大楼坐落于门诊部花园和易北河岸之间，使得病人如同置身于风景之中一般。

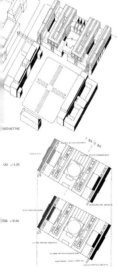

联邦政府办公楼 石勒苏益格—荷尔斯泰因 柏林

竞赛: 1997
设计: Meinhard v. Gerkan mit Gregor Hoheisel
合作者: Hnin Kyaw Lat, Johannes Erdmann

两个省级政府决定在德国首都合作建造，然后合用一个新办公楼。这种合作关系从两座楼反映出来：两翼独立的楼连接着一个公用的大厅，这个大厅还被用作入口大厅，也是通往公园的宽敞通道。

这座建筑在城市规划中据着十分重要的地位，艾伯茨和克莱恩·格兰大街的十字路口上，筑物周围还设定了围合结构。政府官员和嘉宾的房盘踞在大楼的上部，像"祥云"一般。

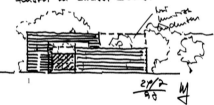

1

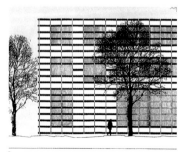

2

3

1 概念草图
2+3 正立面细节
4 透视图
5 总平面图

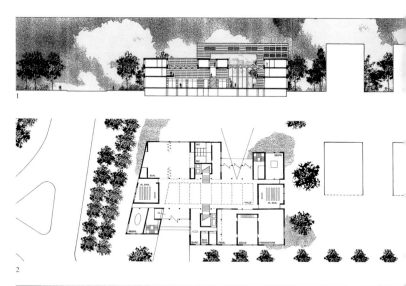

1
2
3

1 纵剖面图
2 一层平面图
3 南立面
4 北立面
5 上层平面图
6 横剖面图和东立面

1 景色规划设计解释了从外部空间到内部的过滤筛选
2 四楼即行政区所在楼层平面图
3 二楼即内部服务区所在楼层平面图
4 地下室平面图
5 模型展示建筑布局
6 总平面图

联邦政府办公楼 勃兰登堡和梅克伦堡—福布莫恩 柏林

竞赛：一等奖，1998
设计：Meinhard v. Gerkan mit Stephan Rewolle
项目负责人：Stephan Rewolle
合作者：Kemal Akay,
Margret Böthig
Antje Pfeifer, Elke Hoffmeister
业主：Land Brandenburg,
Land Mecklenburg-Vorpommern
建造时间：1999-2001
建筑面积：4.430m²
建筑体积：19.290m³

两个同一角度的建筑部分彼此对峙，代表了两个联邦政府。把别墅作为标本加以利用，而整个建筑的统一也清晰可见。由于周围的城市住房向高层建筑发展，作为回应，本建筑在转角处以一个塔楼作标志。中间地带归两个政府共用，是一个多层中庭，中庭上面是玻璃屋顶，用以给各个办公室和其他设施照明。

2

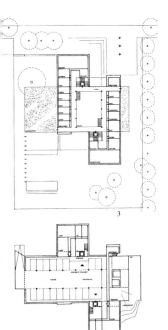

3

4

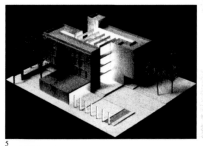

5

6

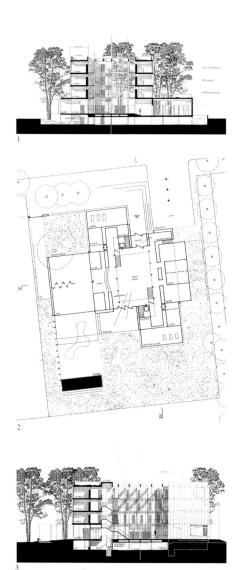

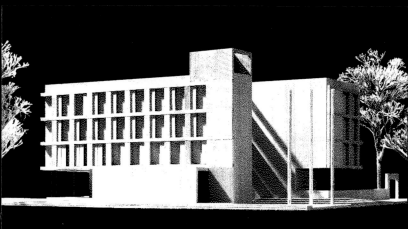

1 A-A 剖面图
2 一层平面图
3 B-B 剖面图
4 着重表现的东北角入口位置
5 西立面图
6 北立面图

阿拉伯联合王国酋长国大使馆住宅 柏林

竞赛：1998
设计：Meinhard v. Gerkan m Johann v. Mansberg

扩建地址就在"万湖"区域最高的地方。建筑中设计了三重中心，融合了要求中的公共空间与私人空间于一体，几乎称得上一个对称体。中央地带营造了一个空间顺序，从入口大厅通过温室可到达餐厅，这里提供了一个直接观看南边湖色的好去处。私人和公共空间都分别被安置在平行的两翼上，两层或两层以上的结构以这种方式分开排列，但总体保持不变。

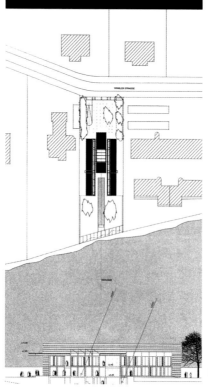

1 模型
2 总平面图及纵剖面图
3 西北立面图。一层平面图。横剖面图。
4 东南立面图。上层平面图。横剖面图。

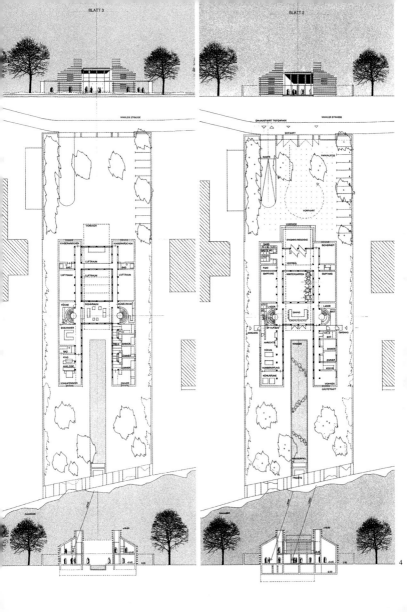

城市规划
URBAN DESIGN

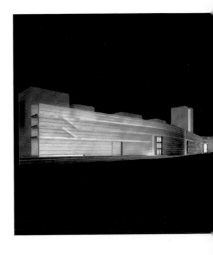

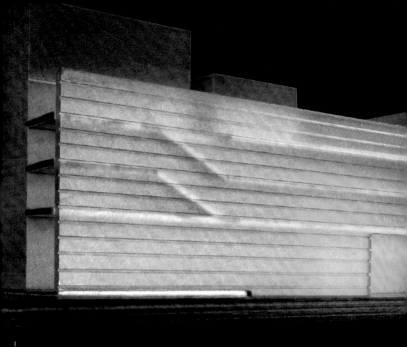

维也纳城市规划新结构

竞赛：二等奖，1996
设计：Meinhard v. Gerkan mit Anja Knobloch-Meding
合作者：Sigrid Müller, Johann v. Mansberg, Stephan Rewolle

规划的目的是给予这个历史性地点新的活力和用途。旧的塔楼作为历史性的参照点及标志性建筑，在整个地区都能看到。新建部分通过沿街而建的玻璃薄膜结构加以界定，使现有的建筑群即使挡在其后也能看见，因而增强了新旧建筑的连接。塔楼前方的中央空地营造出一个新的中心，有影院、成人教育中心和"达勒中心"。最终，新的"达勒塔楼"变成了银幕般的摩天大楼，这十分明显地比喻了旧建筑新的开端，其众多的新功能鼓舞了更深层次的娱乐和探究。

1 临街立面的曲线薄膜结构既是边界又是连接
2-4 整个城市中在这个位置聚焦

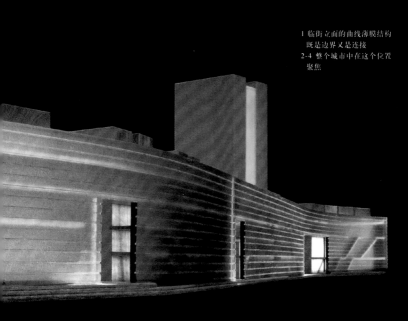

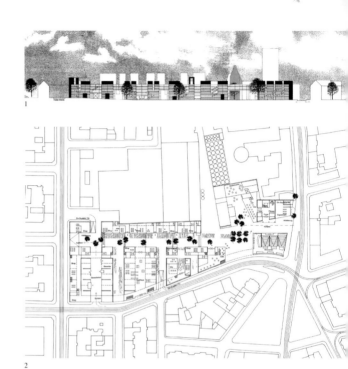
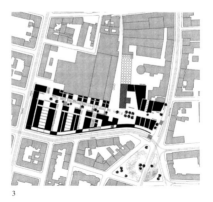

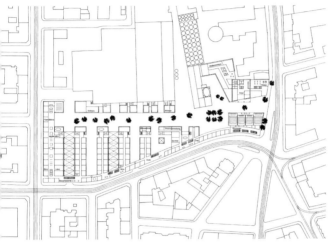

1 南立面
2 一层平面图
3 总平面图
4 东立面
5 上层平面图
6 城镇规划模型

商品交易大楼
巴塞尔

竞赛: 1998
设计: Meinhard v. Gerkan
合作者: Volkmar Sievers,
Brigitte Queck, Bettina Groß,
Sona Kazemi, Sigrid Müller,
Uli Heiwolt

三个新设计的部分用以重新定义交易会大楼前的公共空间: 高高耸起的伴着曲线形正立面的高楼成为标志性建筑——即使在莱茵河对岸也能看到。一个新的建筑物在第三大厅前面被环抱,拥有多层停车场及电车站的波浪状屋顶。公共空间采用统一的铺设,清除了交通和街道设备,使交易会大楼呈现出新的品质,也为来宾提供了更为便捷的方向定位。

1 新的展览塔楼
2 总平面图
3 多层停车场新的正立面

焦油厂 不来梅

竞赛: 1997
设计: Meinhard v. Gerkan
合作者: Charles de Picciotto, Sona Kazemi

此项目在设计中通过简洁的建筑构形和特色鲜明的建筑材料对岛上的环境现状加以呼应。建筑依山形墙而排列，面向一方，而且都悬臂式伸出，接纳两边景色。西北面沿岸的墙群与房屋表现的混凝土的"坚硬感"相匹配。西南立面的建筑饰有金银丝木格子花样，面对新城市广场。

1 整个建筑综合体模型
2 建筑正立面构造

卡伦博格步行街 汉诺威

专家意见: 一等奖, 1991
设计: Meinhard v. Gerkan
项目合伙人: Nikolaus Goetze
项目负责人: Karen Schroeder
合作者: Martina Klostermann,
Michael Haase, Jörg Steinwender,
Cordula v. Graevenitz, Andreas
Perlick
业主: Nileg, Niedersächsische
Gesellschaft für Landesentwicklung
und Wohnungsbau mbH
建造时间: 1996-1999
建筑面积: 44.000m²
建筑体积: 161.823m³

汉诺威市卡伦博格新的商业中心建立了"卡伦博格步行街",其中包括住宅、办公楼、商店以及诊所。设计采用两个梳状结构,两个梳脊相对而立,形成了一个平静的住宅区街道景象。光照充足、修饰美化过的庭院朝向狭窄的街道,如果不是这个设计,街道的环境会很稠密嘈杂。

住宅区林荫大道沿街的树形成一个轴线,通向整个景观一端的纪念碑。朝向卡伦博格路的入口被界定为一座7层高的塔楼,采用了精心构造的轻钢玻璃凸窗结构。住宅区街道两边是高高的拱廊,内设商店和饭馆。

1+2 一个抬高设计的露台成为公寓区的人流通道。入口位于纵向体块建筑下方的小径
3 横剖面图
4 一条小径充当建筑轴线，强化了位于阿道夫街8号的建筑的对称感

1 新旧建筑的结合
2 通向北边的小径充当建筑轴线
3 横剖面图
4 二层平面图

1 临洪保德大街正立面图
2 穿过纵向条块状建筑的梯级通向露台

1 临卡伦博格大街的主楼正立面图
2 圆柱体建筑的标准层平面图
3 建筑正立面采用走廊环绕结构

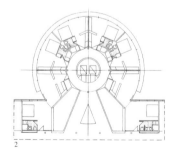

柏林—腓德烈斯汉住宅小区

专家意见: 1992
设计: Volkwin Marg
项目负责人: Sybille Zittlau-Kroos, Dirk Heller
合作者: Asunta Foronda, Ivanka Perkovic, Ruth Scheurer, Susanne Winter
业主: GSW Gemeinnützige Siedungs-und Wohnungsbaugesellschaft Berlin; Grundstücksgesellschaft Friedenstraße Büll & Dr. Liedke
建造时间: 1994-1997
建筑面积: 20.809m²
建筑体积: 63.751m³

沿着腓德烈斯汉大街而建造的住宅成为"腓德烈斯汉社区"项目的第一要素。这个工程对城市设计进行了重新组织并且对勃米斯奇餐厅原址进行了重新设计和征用。设计意在腓德烈斯汉大街和朗斯博格道之间形成封闭的街区。互成角度布局的住宅街区形成了第一部分，还增建了一个新的宾馆建筑。街区分成8个独立的单元，每个单元包含两三种住宅套型，交错的阁楼层参照了传统的柏林式屋顶设计风格。位于地下停车场上方的整个底层空间分配给了零售商，因而建筑底层正立面的处理与楼上的住宅层也有所不同。

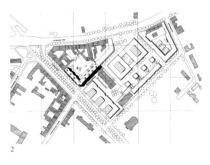

1 面向街区过道的建筑正立面。立面采用立体空间设计，按顺序分别是暖房、凉廊以及阳台
2 总平面图

1 2

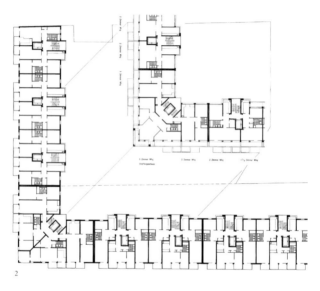

3

+3 街区一角,三维效果图以及实际建筑 标准层平面图及屋顶层平面图

"都市住宅" 柏林

竞赛: 1997
设计: Meinhard v. Gerkan
合作者: Stephan Rewolle

"错层式"住宅正立面采用错开设计,因此可以以多种形态组合,建筑形态轮廓突出而且与街景接触面广。内部空间又可以根据各个住户的住宅概念灵活应用,走廊和剩余面积得到了避免。这种强调垂直感的窄楼占用的建筑面积很小,但是从外部来看,这种楼仍然是清晰可辨的建筑单位。这种楼是按"低耗能住宅"设计的,每平方米消耗的能量不到70kW·h时。

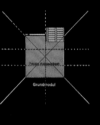

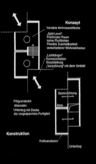

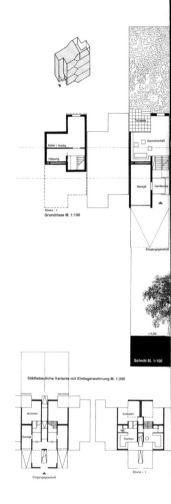

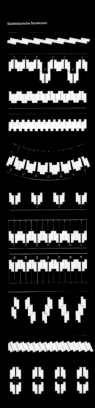

Städtebauliche Strukturen

阿尔斯特弗里特的开发 汉堡

竞赛: 1997
设计: Meinhard v. Gerkan mit Johann v. Mansberg
合作者: Philipp Kamps, Sigrid Müller, Michael Biwer

这个地区的开发填补了最后的城市结构空白，这个空白位于阿尔斯特湖和埃尔伯河之间，沿着狭窄的渠道(小河)轴线，与阿尔斯特湖畔的人行小道平行。这就是对堤岸的公众用途特别强调的原因，如卖古董的拱廊商店、书店和咖啡屋。要求建造的功能空间按一定方式安排，使办公楼面向主大街，而(单个)住宅的花园都坐落在小河畔。

所有的住宅楼都改为三部分建筑。玻璃走廊通向公寓，中间的公寓只面向小河。这种安排提供了不同的住宅套型，从阁楼到小屋都有。

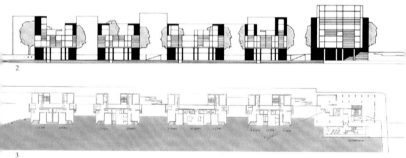

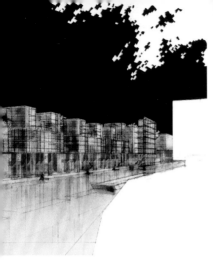
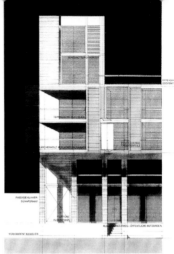

1 建筑南立面 / 立面细部
2 东南立面图
3 入口层平面图
4 横剖面图
5 在整体城市环境中的位置

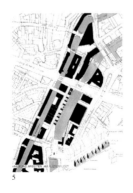

建筑学院　杜塞尔多夫

竞赛: 1998
设计: Meinhard v. Gerkan mit Nicolai Pix

基地所处位置与莱茵河平行，使得设计寓意了一艘"建筑师的船"——几百年来这都是与建筑联系在一起的象征。体现"船"的寓意的是建筑底层不设窗户的设计，以及上面的五个"货房"，与吃水线垂直、中间有小港湾间隔。五个楼翼都是通过天桥连接，每一个建筑都因在屋顶层缩进两个台阶而显得造型独特。

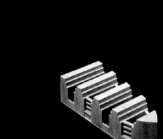

1+2 模型
3 立面图
4 一层平面图
5 标准层平面图
6 纵剖面图

贝内迪克斯广场商住楼 埃尔富特

竞赛：1997
设计：Meinhard v. Gerkan mit Philipp Kamps

设计通过这一底层拱形过道将这个办公及住宅楼与菲施马克特广场、本蒂克布拉茨广场、以及通往前犹太教会堂的公众通道连接起来，作为整个城市规划的一部分。这种创意通过简单的封闭式拐角建筑得以实现，旨在维持城市结构的完整形态。

建筑在垂直方向上分三段设计：首先在底层采用混凝土石块，然后是主立面采用砂岩材料、并镶嵌建筑专用铜材。整个建筑物上覆盖的是梯式屋顶。

1 模型研究
2 西立面图、总平面图
3 剖面图、平面图
4 立面细部

2000年世界博览会汉诺威

可行性研究: 1998
设计: Meinhard v. Gerkan
Joachim Zais
合作者: Monika van Vught, Cornelia Thamm, Gabriele Wysocki,
Ulf Düsterhöft, Hilke Eustrup, Rolf Duerre, Olaf Schlüter, Thomas Dreusicke, Helge Reimer, Thomas Kehl
委托人: Expo 2000 Hannover GmbH

研究表明,在世界博览中心广场的边界区域进行临时开发是可行的。如果投资开发未能完全实现的话,展览会期间建设的建筑就只能为博览会服务。设计建议采用轻钢玻璃结构以及高度灵活的平面布局。在两层高的建筑围绕庭院并在延伸处增建了塔楼。塔楼的高度达到了指定的18m,所以保持了一定的空间比例。世博会的旗帜可以在这些塔楼上升起。结构框架在水平和垂直方向上基于同一中心轴线。庭院采用轻型钢材篷顶材料,因此围合出一些高9m,室内没有支柱的多功能厅。楼梯安置在各个拐角,而外部电梯确保了清晰的垂直交通组织。庭院中建议使用了柱廊,朝向中心开放。起承重作用的钢结构全部经过电镀,立面材料采用防雨的"贝托普朗"材料或螺纹铝材。遮阳天窗也同样采用电镀钢板材料。

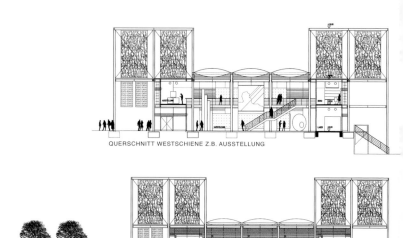

ANSICHT STRASSE WESTSCHIENE TYP E Z.B. FOOD-COURT

LÄNGSSCHNITT WESTSCHIENE Z.B. FOOD-COURT

ANSICHT SÜDEN OSTSCHIENE TYP B Z.B. EINGANG OST/MERCHANDISING

LÄNGSSCHNITT OSTSCHIENE Z.B. EINGANG OST/MERCHANDISING

QUERSCHNITT OSTSCHIENE Z.B. EINGANG OST/MERCHANDISING

ANSICHT PLAZA OSTSCHIENE TYP A Z.B. EINGANG OST/MERCHANDISING

ANSICHT STRASSE WESTSCHIENE TYP D Z.B. GEFLÜGELHAUS

LÄNGSSCHNITT WESTSCHIENE Z.B. INTERNATIONALE ORGANISATION

QUERSCHNITT WESTSCHIENE Z.B. GEFLÜGELHAUS

ANSICHT PLAZA WESTSCHIENE TYP A Z.B. GEFLÜGELHAUS

CHNITT

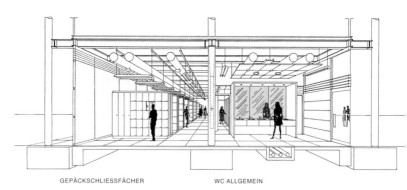

GEPÄCKSCHLIESSFÄCHER   WC ALLGEMEIN

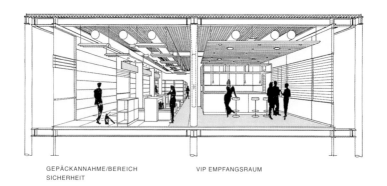

GEPÄCKANNAHME/BEREICH
SICHERHEIT   VIP EMPFANGSRAUM

VARIANTE "GEFLÜGELHAUS" GRUNDRISS OG

VARIANTE "BISTRO/CONVENIENCE STORE" GRUNDRISS OG

VARIANTE "FOOD-COURT" GRUNDRISS OG

VARIANTE "GEFLÜGELHAUS" GRUNDRISS EG

VARIANTE "BISTRO/CONVENIENCE STORE" GRUNDRISS EG

VARIANTE "FOOD-COURT" GRUNDRISS EG

VARIANTE "AUSSTELLUNG" GRUNDRISS OG

VARIANTE "INTERNATIONALE ORGANISATION" GRUNDRISS OG

VARIANTE "EINGANG OST/ MERCHANDISING" GRUNDRISS OG

VARIANTE "AUSSTELLUNG" GRUNDRISS EG

VARIANTE "INTERNATIONALE ORGANISATION" GRUNDRISS EG

VARIANTE "EINGANG OST/ MERCHANDISING" GRUNDRISS EG

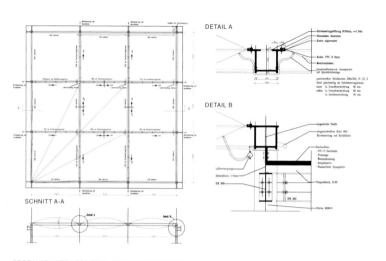

DETAIL A

DETAIL B

SCHNITT A-A

PROFILVARIANTEN DER STAHL TRAGKONSTRUKTION-
TRÄGERANSCHLÜSSE AN STÜTZEN

VARIANTE A

VARIANTE B

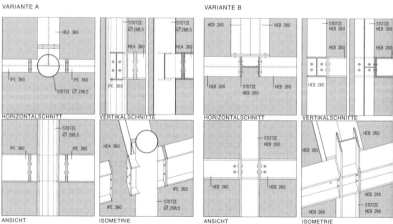

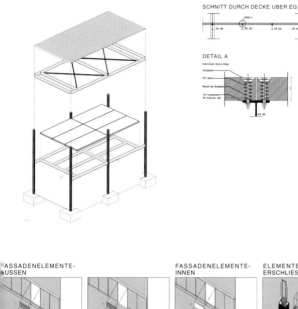

SCHNITT DURCH DECKE ÜBER EG

DETAIL A

VERTIKALSCHNITT STAHLBAUKNOTEN

HORIZONTALSCHNITT STAHLBAUKNOTEN

FASSADENELEMENTE-AUSSEN

FASSADENELEMENTE-INNEN

ELEMENTE DER ERSCHLIESSUNG

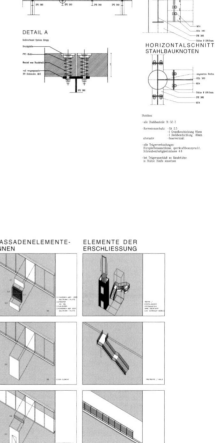

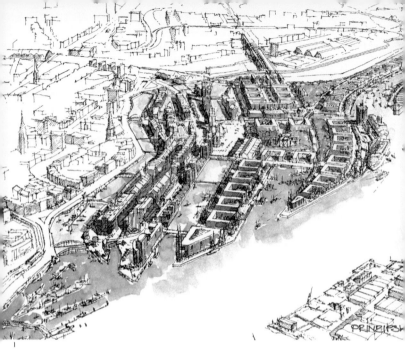

1 建筑标志的鸟瞰图
2 1996 年港口地区的鸟瞰照片

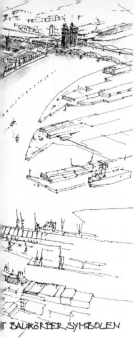

港口城市　汉堡

可行性研究: 1997
设计: Volkwin Marg
修改: RWTH Aachen Fakultät für Architektur, Lehrstuhl für Stadtbereichsplanung
委托人: Gesellschaft für Hafen- und Standortentwicklung GHS

在北约和欧盟对东欧开放并重新统一之后,传统意义上被称为内地的拥有1.5亿人口的汉堡的经济潜力得到再次发展——这是一个不同寻常的经济契机。这个过程要求更多的城市服务区,这远非现有的城市中心所能提供。在开发的过程中,重建的港口成为靠近内城的地区,并且与具有历史意义的"斯贝彻城"相毗邻。"斯贝彻城"被定义为新的"港口城",而且将其进行了多种功能的可行性研究。

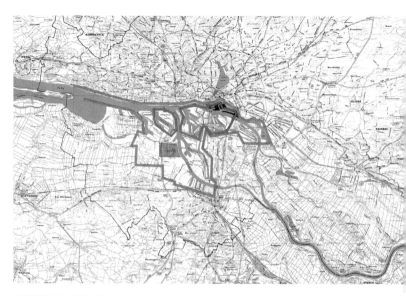

1 规划区域的中心和周边区域
2 1996年原有建筑的平面结构形态
3 设计方案的平面结构形态

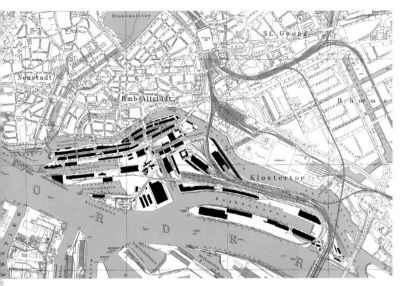
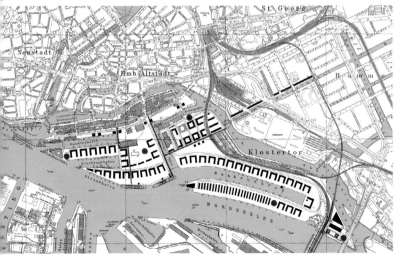

隔声墙　累根斯堡

竞赛: 1998
设计: Meinhard v. Gerkan mit Sona Kazemi

伯格魏提格地区上的艺术概念就是一个极富想像力的百万平方米的四边形。把这个地区划分为四个相等的正方形，使其四个区域都能展现自己特点：

——绿地；
——村庄类型的建筑平面结构形态；
——自成一体的住宅区；
——周界区开发。

隔声墙也是整体概念的一部分。它沿着铁道，不仅是隔声墙，也是一个屏幕。隔声墙只采用两种表面材料：木制平板和透明纹理的玻璃，玻璃之间只空出垂直罅隙。

这些都按变动间隔排列造成了"条形码"那种变化的形象，变化视列车的速度而定，因此造成了时间流逝的效果，使得我们不难想到商业社会快速的运行机制。

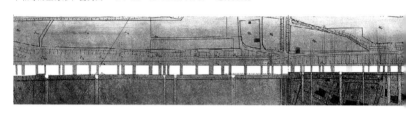

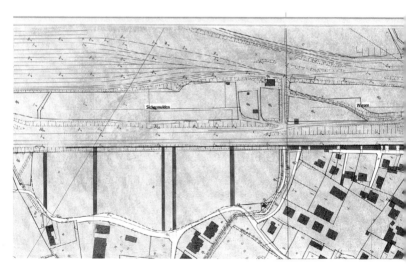

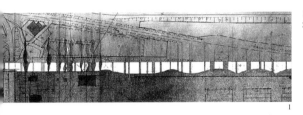

1 隔声墙沿线的布局顺序
2 总平面图及周边公共空间

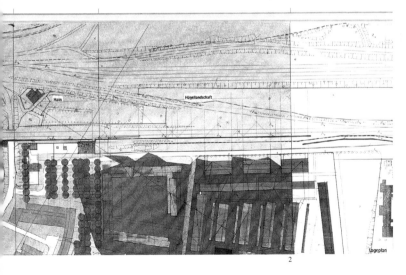

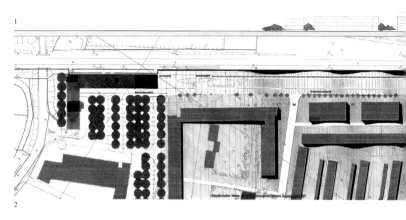

1 剖面图
2 隔声墙沿线的开放空间概念
3 渗流渠
4 草地
5 小树林
6 山景
7 从保护区方向看的建筑立面
8 临街立面图
9 细部处理

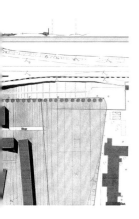
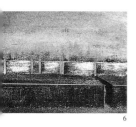
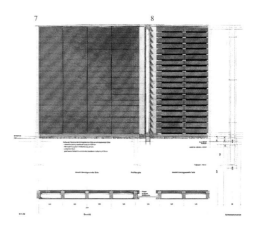
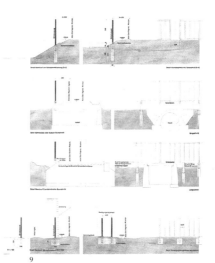

住宅 舍内贝克

设计: Volkwin Marg; 1995
项目负责人: Martin Bleckmann
合作者: Jutta Hartmann-Pohl, Franz Lensing, Olaf Drehsen, Monika Kaesler
静力分析: Ing.-Büro Kempen
建筑技术: HL-Technik
景观设计: WES und Partner
业主: DSR Immobilien
建造时间: 1996-1997
建筑面积: 6.350m²
建筑体积: 23.600m³

舍内贝克临近柏林，是一块小型地产，地处一片树林之中，开发为一个开放式住宅，包括一些一层或两层的楼房、醒目的林荫道和大片的绿地。迄今为止，将一个建筑中心作为城镇的中心定位这种设计方式已经过时。参照当地传统建造布局，设计完成了带有一条南北向小径的南部中心要求有一条南北方向的小径，另外还有一些两三层的住宅。工程要一分为二，包括四个单体。每座楼都要有屋顶花园，而底楼用作各种大小的商店。花园区对于公寓来说相当于流通部位。朝向街道和广场有一个拱形建筑坐落在商业区的前面，形成城市公共区域的一部分，在停车区还有钢建筑的凉亭横跨而过。这与修饰过的花园相结合，再加上整体的座椅设计和植被排列，增强了这个地点的"中心"特点。

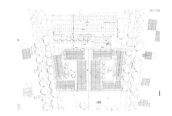

1 总平面
2 综合体规划方案
3 屋顶层立面结构

1 通过屋顶层的外部走廊可以从流通区域进入公寓
2+3 南面视角。中轴线后面就是拱形建筑
4 从西北方向看向西面街区

修复和室内设计
RESTORATION AND INTERIOR DESIGN

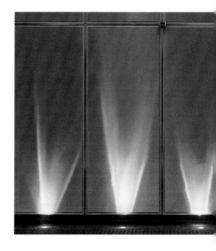

塞勒大剧院的重修
汉堡

设计: Klaus Staratzke
项目负责人: Dagmar Winter
合作者: Maja Gorges
业主: Thalia Theater
建造时间: Juni-okt. 1997

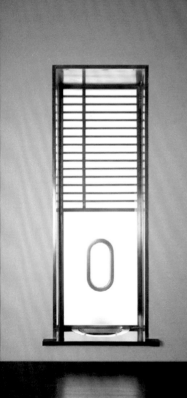

这个剧院最初是由乔治·卡尔摩根在1912年建造的,在战争期间遭到一定破坏,然后于1960年由其子华纳加以改建。内部各种建筑类型反映了各个不同时期的建筑风格,因此在重建过程中必须谨慎考虑建筑保护方面的要求。所以,售票厅内古旧的彩花地板以及入口两层楼高的木制护壁在老卡尔摩根的要求下得以保留。20世纪60年代的环绕走廊按照衣帽间的纵向顺序进行了空间定义。木料的颜色配置得到部分的保留,大理石和黄铜被去除,取而代之的是小卡尔摩根的白色漆面、木料和着色地毯相

1 售票口
2+3 改建前后的入口前厅

1

2

1+2 改建前后的前厅
3 衣帽间的桌子
4 门廊。改建工程清楚了原有许多设计元素。美观的磨石地被重新暴露出来
5+6 地毯及过于耀眼的照明设计破坏了环境氛围

哈帕格·罗伊德公司总部 改建与修缮
巴林顿 汉堡

竞赛: 1997
设计: Volkwin Marg,
Klaus Staratzke
项目负责人: Dagmar Winter
合作者: Kerstin Steinfatt, Maja
Gorges
业主: Hapag Lloyd AG
建造时间: Jan.-Juni 1997

为庆祝其150年大典,哈帕格·罗伊德要求对公司总部进行修缮工作并尽可能地对华丽的入口直至总部区进行复原,这些是在1912-1923年由福里茨·霍格尔作为原建筑的扩建而造的,原建筑由马丁·霍勒尔(1901-1903年)建造。在考虑保留的同时,也应该顺应公司发展的现代化要求。入口处采用对称设计,按休息室、小厅和主厅的顺序复原。玻璃屋顶被清除后重新按照原貌铺设新的玻璃材料。旧的巨大"通道大厅"是对美国举办交易会用的,只能通过减少大量的内置设施来复原。

大厅现在被用作可分刷的会议厅,配有现代计算机技术装置。一切现有的材料都经过精心地复原,而整块大理石地板则刷洗一新。

1 复原的玻璃天窗。屋顶结构采用新铺设的轻钢玻璃结构
2 复原的顶棚结构:漆木上的镀金镶嵌
3 重建后的主厅及前厅
4 重建之前的主厅

1 根据使用要求,大厅可以通过人造照明来调节光线明暗
2 重建之前的会议室使用吊顶,没有天光
3 重建了玻璃天窗的会议室
4+5 小会议室:或可以眺望阿尔斯特湖景色,或可以看到餐具间,餐具间用一扇镶嵌玻璃的木移门隔开

1 会议室照明墙采用不锈钢网板
2 依照声学原理的墙面铺细节:采用木质框架的不锈钢网板

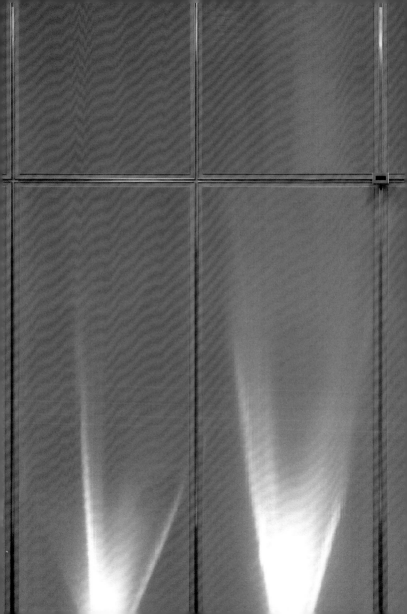

戏剧院休息室大厅和衣帽间的修缮

竞赛：三等奖，1997
设计：Meinhard v. Gerkan
合作者：Michael Biwer, Stephan Rewolle

戏剧院的休息室和衣帽间看起来毫无特色，与现代的要求不相符。设计连接了两个分开的楼层，通过引入可移动屏幕覆盖在现有的建筑部位上，创造了一种新的空间上的和谐。新的空间印象通过延续体积来实现，增强了主要流通楼梯下的活动，并充分利用其可利用的空间。

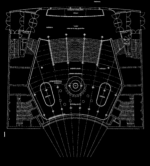

1 入口层平面图
2 地下室平面图
3 剖面图
4 可移动屏幕同时连接了两个楼层，以形成一个统一体
5 衣帽间内视

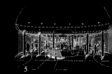

附录
APPENDIX

MEINHARD VON GERKAN
PROF. DIPL. -ING. ARCHITEKT BDA

geboren am 3. Januar 1935 in Riga/Lettland.
1964 Diplom-Examen an der TU Braunschweig.
seit
1965 Freiberuflicher Architekt zusammen mit Volkwin Marg.
1972 Berufung in die Freie Akademie der Künste in Hamburg.
1974 Berufung an die TU Braunschweig als ordentlicher Professor/Lehrstuhl A für Entwerfen.
1982 Berufung in das Kuratorium der Jürgen-Ponto-Stiftung, Frankfurt.
1965 -
1995 Mehr als 300 Preise in nationalen und internationalen Wettbewerben, darunter mehr als 130 1. Preise zusammen mit Volkwin Marg. Zahlreiche Preise für vorbildliche Bauten. Zahlreiche Veröffentlichungen im In- und Ausland. Zahlreiche Preisrichter- und Gutachtertätigkeit.
1988 Gastprofessor an der Nihon Universität, Tokio/Japan.
1993 Gastprofessor an der University of Pretoria/Südafrika.
1995 American Institute of Architects, Honorary Fellow, USA. Ehrenauszeichnung der Mexikanischen Architektenkammer.

born on 3 January 1935 in Riga/Latvia.
1964 Diploma examination at the TU Braunschweig.
since
1965 Free lance architect together with Volkwin Marg.
1972 Appointment to the Freie Akademie der Künste in Hamburg.
1974 Appointment to the Technische Universität Braunschweig as professor/course A for design.
1982 Appointment to the board of the Jürgen-Ponto-Foundation, Frankfurt.
1965 -
1995 More than 300 national and international competition prizes incl. more than 130 1st prizes together with Volkwin Marg. Many awards for outstanding buildings. Many publications in Germany and abroad. Considerable involvement with competition juries and reports.
1988 Guest professor at the Nihon University, Tokyo/Japan.
1993 Guest professor at the University of Pretoria/South Africa.
1995 American Institute of Architects, Honorary Fellow, USA. Honored by the Mexican Architectural Society.

VOLKWIN MARG
PROF. DIPL. -ING. ARCHITEKT BDA

geboren am 15. Oktober 1936 in Königsberg/Ostpreußen, aufgewachsen in Danzig.
1964 Diplom-Examen an der TU Braunschweig.
seit
1965 Freiberuflicher Architekt mit Meinhard v. Gerkan. Zahlreiche Wettbewerbserfolge und große Bauaufträge. Vorträge und Texte zu Fragen der Architektur, des Städtebaus und der Kulturpolitik.
1972 Berufung in die Freie Akademie der Künste in Hamburg.
1974 Berufung in die Deutsche Akademie für Städtebau und Landesplanung.
1975 -
1979 Vizepräsident des Bundes Deutscher Architekten BDA.
1979 -
1983 Präsident des BDA.
1986 Berufung an die RWTH Aachen, Lehrstuhl für Stadtbereichsplanung und Werklehre.
1996 Fritz-Schumacher-Preis der FVS-Stiftung.

born on 15 October 1936 in Königsberg/Ostpreußen, childhood in Danzig.
1964 Diploma examination at the TU Braunschweig.
since
1965 Free lance architect together with Meinhard von Gerkan. Many competition successes and large projects, lectures and manuscripts on architecture, urban planning and political culture.
1972 Appointment to Freie Akademie der Künste Hamburg.
1974 Appointment to Deutsche Akademie für Städtebau und Landesplanung.
1975 -
1979 Vice President of the Bund Deutscher Architekten BDA.
1979 -
1983 President of the Bund Deutscher Architekten BDA.
1986 Appointment to the Chair of Town Planning and Tradesmanship, RWTH Aachen.
1996 Fritz-Schumacher-Award.

KLAUS STA RATZKE
DIPL. -ING. ARCHITEKT

geboren am 12. Dezember 1937 in Königsberg/Ostpreußen.
1963 Diplom-Examen an der TU Berlin.
1963 -
1966 Freier Mitarbeiter im Architekturbüro Hentrich + Petschnigg, Düsseldorf.
1968 Mitarbeit im Büro von Gerkan und Marg.
1972 Partner im Büro von Gerkan, Marg und Partner.

born on 12 December 1937 in Königsberg/Ostpreußen.
1963 Diploma examination at the TU Berlin.
1963 -
1966 Free lance work by Architekturbüro Hentrich + Petschnigg, Düsseldorf.
1968 Work with von Gerkan and Marg.
1972 Partner of von Gerkan, Marg and Partners.

UWE GRAHL
DIPL. -ING. (FH) ARCHITEKT AIV

geboren am 19. Oktober 1940 in Dresden.
1959 Maurerlehre, Gesellenbrief.
1963 Examen an der Staatlichen
Ingenieurschule für
Bauwesen Berlin-Hochbau.
1963 Mitarbeit im Büro
Dipl.-Ing. Siegfried Fehr.
1969 Mitarbeit im Büro
Dipl.-Ing. Rolf Niedballa.
seit
1974 Büro von Gerkan, Marg und Partner,
Berlin.
seit
1990 Assoziierter Partner im Büro
von Gerkan, Marg und Partner.
seit
1993 Partner im Büro von Gerkan, Marg
und Partner.

born on 19 October 1940 in Dresden.
1959 Bricklayer apprentice, trade
certificate.
1963 Examination at the
Staatliche
Ingenieurschule für Bauwesen
1963 Work with Büro
Dipl.-Ing. Siegfried Fehr.
1969 Work with Büro
Dipl.-Ing. Rolf Niedballa.
since
1974 Office von Gerkan, Marg
and Partners, Berlin.
since
1990 Associate partner of
von Gerkan, Marg and Partners.
since
1993 Partner of von Gerkan, Marg
and Partners.

JOACHIM ZAIS
DIPL. -ING. ARCHITEKT BDA

geboren am 10. Dezember 1951 in
Marburg/Lahn.
1969 Tischlerlehre in Hildesheim.
1975 Examen an der FH Hildesheim mit
Abschluß Ing. grad.
1975 Architekturstudium an der
TU Braunschweig.
Tätigkeit während des Studiums
in verschiednen Architektur-
büros und Wettbewerbstätigkeit.
1982 Diplom an der TU Braunschweig.
1982 Tätigkeit im Büro für
Stadtplanung Dr. Schwerdt,
Braunschweig.
1983 –
1989 Assistententätigkeit am Institut
für Baugestaltung A – Prof. M. v.
Gerkan.
Freier Mitarbeiter im Büro
von Gerkan, Marg und Partner,
Braunschweig,
und eigene Tätigkeit als
Architekt.
1989 Leitung des Büros
von Gerkan, Marg und Partner,
Braunschweig.
seit
1993 Partner im Büro
von Gerkan, Marg und Partner.

born on 10 December 1951 in
Marburg/Lahn.
1969 Carpenter apprentice in
Hildesheim.
1975 Examination Ing. grad. at the
FH Hildesheim.
1975 Architectural studies at the
TU Braunschweig.
Worked during studies for
different architects and
on various competitions.
1982 Diploma at the TU Braunschweig.
1982 Worked in Büro für Stadtplanung
Dr. Schwerdt, Braunschweig.
1983 –
1989 Lecturer at Institut für Baugestaltung A – Prof. M.v. Gerkan.
Free lance architect with
von Gerkan, Marg and Partners and
as architect on own projects.
1989 Head of office von Gerkan, Marg
and Partners, Braunschweig.
since
1993 Partner of von Gerkan, Marg
and Partners.

HUBERT NIENHOFF
DIPL. -ING. ARCHITEKT

geboren am 4. August 1959 in Kirchhellen/
Westfalen.
1985 Diplom-Examen an der
RWTH Aachen.
1985 –
1987 Mitarbeit im Büro für
Architektur und Stadtbereichsplanung
– Ch.Mäckler, Frankfurt/Main.
1987 –
1988 Auslandsaufenthalt in den USA,
städtebauliche Studien.
1988 –
1991 Assistent an der RWTH Aachen,
Lehrstuhl für Stadtbereichsplanung und Werklehre.
Prof. Volkwin Marg.
1988 Mitarbeit im Büro
von Gerkan, Marg und Partner, Aachen.
seit
1993 Partner im Büro
von Gerkan, Marg und Partner.

born on 4 August 1959 in Kirchhellen/West-
falen.
1985 Diploma examination at the
RWTH Aachen.
1985 –
1987 Work with Büro für Architektur
und Stadtplanung – Ch.Mäckler,
Frankfurt/Main.
1987 –
1988 Foreign visit to USA with urban
studies.
1988 –
1991 Lecturer at the RWTH Aachen
Chair for Stadtbereichsplanung
und Werklehre.
Prof. Volkwin Marg.
1988 Work with von Gerkan, Marg
and Partners.
since
1993 Partner of von Gerkan, Marg
and Partners.

WOLFAGNG HAUX
DIPL. -ING. ARCHITEKT BDA

geboren am 13. August 1947 in Hamburg.
1969 Architekturstudium an der Hochschule für Bildende Künste.
1975 Diplom-Examen.
1976 Mitarbeit im Architekturbüro Prof. Dieter Hoor, Steinhorst.
seit
1978 Mitarbeit im Büro von Gerkan, Marg und Partner, Hamburg.
seit
1994 Assoziierter Partner im Büro von Gerkan, Marg und Partner.

born on 13 August 1947 in Hamburg.
1969 Architectural Studies at Hochschule für Bildende Künste.
1975 Diploma Examination.
1976 Work with Architekturbüro Prof. Dieter Hoor, Steinhorst.
since
1978 Work with von Gerkan, Marg and Partners, Hamburg.
since
1994 Associate Partner with von Gerkan, Marg and Partners.

NIKOLAUS GOETZE
DIPL. -ING. ARCHITEKT

geboren am 25. September 1958 in Kempen.
1980 Architekturstudium an der RWTH Aachen.
1985 –
1986 Meisterklasse Prof. W. Holzbauer, Hochschule für angewandte Kunst, Wien.
1987 Diplom an der RWTH Aachen.
seit
1987 Mitarbeit im Büro von Gerkan, Marg und Partner, Hamburg.
1994 Assoziierter Partner im Büro von Gerkan, Marg und Partner.
seit
1998 Partner im Büro von Gerkan, Marg und Partner.

born on 25 September 1958 in Kempen.
1980 Architectural studies at RWTH Aachen.
1985 –
1986 Master Class Prof. W. Holzbauer Hochschule für angewandte Kunst, Wien.
1987 Diploma at the RWTH Aachen.
since
1987 Work with von Gerkan, Marg and Partners.
1994 Associate Partner with von Gerkan, Marg and Partners.
since
1998 Partner of von Gerkan, Marg and Partners.

JÜRGEN HILLMER
DIPL. -ING. ARCHITEKT

geboren am 26. Dezember 1959 in Mönchengladbach.
1980 Architekturstudium an der Carolo-Wilhelmina in Braunschweig.
1988 Diplom.
1988 –
1992 Mitarbeit im Büro von Gerkan, Marg und Partner, Hamburg.
1992 –
1995 freiberuflicher Architekt in Haltern, Nordrhein-Westfalen.
1994 Assoziierter Partner im Büro von Gerkan, Marg und Partner.
seit
1998 Partner im Büro von Gerkan, Marg und Partner.

born on 26 December 1959 in Mönchengladbach.
1980 Architectural studies at the Carolo- Wilhelmina in Braunschweig.
1988 Diploma.
1988 –
1992 Work with von Gerkan, Marg and Partners, Hamburg.
1992 –
1995 Free lance architect in Haltern, Nordrhein-Westfalen.
1994 Associate Partner with von Gerkan, Marg and Partners.
since
1998 Partner of von Gerkan, Marg and Partners

功的竞赛参与

等奖

'64	1.	Sports and Conference Hall, Hamburg
	2.	Indoor and open-air swimming pool, Brunswick
'65	3.	Indoor and open-air swimming pool, SPD
	4.	Max Planck Institute, Lindau/Harz
	5.	Tax Bureau, Oldenburg
	6.	Sports Center, Diekirch/Luxemburg
	7.	Airport Berlin-Tegel
'66	8.	Stormarn Hall, Bad Oldesloe
	9.	District indoor swimming pool, Cologne
	10.	Sports Forum, University of Kiel
'70	11.	Shell AG head offices, Hamburg
	12.	District Vocational School, Bad Oldesloe
'71	13.	European Patent and Trademark Office, Munich
	14.	Multi-purpose building III, University of Hamburg
	15.	Housing Gellertstrasse, Hamburg
	16.	Shopping mall, Alstertal, Hamburg
'72	17.	ARAL AG head offices, Bochum
	18.	Schools Center, Friedrichstadt
'74	19.	Vocational Training Center G 13, Hamburg-Bergedorf
	20.	Provinzial Insurance offices, Kiel
'75	21.	Deutscher Ring, Cologne
	22.	Airport, Munich II
'76	23.	District Administration, Recklinghausen
	24.	Airport, Moscow
	25.	Airport, Algiers
'77	26.	Parish House, Suhl
	27.	Otto mail-order company, head offices extension, Hamburg
	28.	MAK head offices, Kiel
	29.	Police Station, Pankstrasse, Berlin
'78	30.	Pahlavi National Library, Tehran
	31.	Joachimsthaler Platz, Berlin
	32.	Federal German Ministry of Transportation, Bonn
'79	33.	Indoor/outdoor swimming pool, Berlin-Spandau
	34.	Indoor swimming pool for competitions, Mannheim-Herzogenried
	35.	Institute of Chemistry, University of Brunswick
	36.	Institute of Biochemistry, University of Brunswick
	37.	Vereinsbank, Hamburg
	38.	District Administration, Meppen
'80	39.	Academy of Fine Arts, Hamburg
	40.	Römerberg development Frankfurt/Main – free design
	41.	Fleetinsel development, Hamburg
	42.	Lazarus Hospital, Berlin
	43.	Trade and Vocational Training Center, Flensburg
	44.	Deutsche Lufthansa office building, Hamburg
	45.	Airport, Stuttgart
	46.	Johanneum 1 sports hall, Lübeck
'81	47.	Civic Center, Bielefeld
	48.	Kravag offices, Hamburg
	49.	Hotel Plaza, Bremen
	50.	DAL office center, Mainz
	51.	Housing complex, Bad Schwartau
	52.	Law Court, Hamburg
	53.	Refurbishment/conversion, Kiel Castle
	54.	'Komplex Rose', hospital for rheumatic diseases, Bad Meinberg
1983	55.	Gruner + Jahr publishing house, Hamburg
1984	56.	Quickborn block corner development
1985	57.	District Court, Flensburg
	58.	Museum and library, Münster
	59.	Interior restructuring, Bertelsmann, Gütersloh
1986	60.	Town Hall, Husum
	61.	Airport, Hamburg-Fuhlsbüttel
	62.	Bäckerstrasse, Halstenbek
	63.	Labor Office, Oldenburg
	64.	Zürich-Haus, Hamburg
1988	65.	Federal Ministry of Environmental Protection and Reactor Safety, Bonn
	66.	Salamander building, Berlin
	67.	Störgang, Itzehoe
	68.	EAM, Kassel
	69.	Station square, Koblenz
1989	70.	Residential Park Falkenstein, Hamburg
	71.	Carl Bertelsmann Foundation, Gütersloh
1990	72.	Music and Conference Hall, Lübeck
	73.	Deutsche Revision, Frankfurt/Main
	74.	Adult Educational Institute and Municipal library, Heilbronn
	75.	Landscaped apartment complex, Kandestrasse, Hamburg-Nienstedten
	76.	Technology Center, Munster
1991	77.	German-Japanese Center, Hamburg
	78.	Hansetor, Hamburg-Bahrenfeld
	79.	Shopping mall, Langenhorn Markt, Hamburg
	80.	Office complex, Am Zeppelinstrasse, Bad Homburg
	81.	Lenné-Passage, Frankfurt/Oder
	82.	Altmarkt Dresden, 2nd stage
1992	83.	New Trade Fair, Leipzig
	84.	Forum Neukölln, Berlin
	85.	Allee-Center, Leipzig-Grünau
	86.	Commercial building and library, Mollistrasse, Berlin
	87.	Lecture Theater Center, University of Oldenburg
	88.	Siemens Nixdorf, Service Center, Munich
	89.	District Court North, Hamburg
	90.	Apartment block, Schöne Aussicht, Hamburg
	91.	Telekom operating building, Suhl
	92.	Mare Balticum Hotel, Bansin/Usedom
1993	93.	Lehrter Train Station in Berlin
	94.	German-Japanese Center, Berlin
	95.	Office building, Bredeney, Essen
	96.	Labor Office Training Center of the Federal Republic of Germany, Schwerin
	97.	Trade Fair, Hanover, Hall 4
1994	98.	Forum Koepenick, Berlin
	99.	Gerling Insurance Corp., residential and office villas, Leipzig
	100.	Lecture Theater Center, extension of Chemnitz Technical University
	101.	Building for cooperative Norddeutsche Metall-Berufsgenossenschaft, Hanover
	102.	New Center, Berlin-Schönefeld
1995	103.	Neumarkt, Celle
	104.	Dresdner Bank, Pariser Platz, Berlin
	105.	Potsdam Center, southern site (main station Potsdam Stadt)
1996	106.	Hamburg Agency in Berlin, Palais Lusenstrasse
	107.	Civic Center, Weimar
	108.	Apartment and commercial building Deutrichshof, Leipzig
	109.	Expo 2000 Plaza, Hanover
	110.	Bucharest 2000
	111.	Altmarkt Dresden, south-western side
1997	112.	Facade design, Train Station Potsdam-Stadt
	113.	Trade Fair, Hanover, Hall 8/9
	114.	Footbridges, Expo 2000
	115.	New Station Neighborhood, Bielefeld
	116.	Media-Center Leipzig
	117.	IGA 2003 Rostock
	118.	Trade Fair, Rimini
	119.	Weserbahnhof II, Grothe Museum, Bremen
	120.	Casino, Bad Steben
	121.	Christian Pavilion, Expo 2000
	122.	Trade Fair, Düsseldorf
	123.	Builders' training yard, Berlin-Mahrzahn
1998	124.	Stuttgart Airport, terminal 3
	125.	German School and Service Housing in Beijing, China
	126.	Agency of Sachsen-Anhalt in Berlin
	127.	Agency of Brandenburg and Mecklenburg-Vorpommern in Berlin
	128.	Art Kite Museum, Detmold
	129.	Berlin-Brandenburg International Airport
	130.	Philips Convention Stand
	131.	Berlin Olympic Stadium, conversion
1999	132.	Nanning Convention & Exhibition Center, China
	133.	Schloß Hopferau
	134.	Tempodrom, Berlin
	135.	Ancona Airport, Italy

二等奖

1959	1.	District Administration, Niebüll
1966	2.	Jungfernstieg, Hamburg
1967	3.	Buildings for the 20th Olympic Games, Munich, project B
1968	4.	Church Center, Hamburg-Ohlsdorf
1970	5.	Schools Center, Heide-Ost
	6.	Regional Postal Directorate, Bremen
1971	7.	Regional Tax Bureau, Hamburg
	8.	Sports Forum, University of Bremen
1972	9.	Regional Administration offices, Lüneburg
1975	10.	Housing Billwerder-Allermöhe, Hamburg
	11.	Regional Department of the Interior, Kiel
1976	12.	Urban design Universität-Ost, Bremen
	13.	Administrative offices, operating area of Airport Munich II
1977	14.	Row houses and urban villas, Hamburg Bau 78
1980	15.	Gasworks, Munich
	16.	Administrative offices for Volkswagen, Wolfsburg
	17.	Town Hall, Oldenburg
	18.	Max Planck Institute for Quantum Optics, Munich
1981	19.	Institute of Chemistry, University of Brunswick
	20.	Labor Office, Kiel-Hörn
	21.	Klinikum II, Nuremberg-South
	22.	Kravag offices, Hamburg
1982	23.	Schlossparkhotel 'Orangerie', Fulda
1983	24.	Daimler Benz AG offices, Stuttgart
1984	25.	Germanisches Nationalmuseum, Nuremberg
	26.	Intermediate School and Sports Hall, Schleswig-Holstein

335

1985	27. Wildlife Information Pavilion Balje	1980	12. University of Bremerhaven
1986	28. Post offices 1 and 3, Hamburg	1985	13. Urban design competition, Münster
	29. Kummellstrasse, Hamburg	1986	14. Bundeskunsthalle, Bonn
	30. Technik III, Gesamthochschule Kassel	1987	15. Police Headquarters, Berlin
	31. Dockland development, Heiligenhafen	1988	16. Library of Technical University and Fine Arts Academy, Berlin
1987	32. Telecommunications towers for German Federal postal services	1989	17. Concert Hall, Dortmund
	33. Research and lecture hall building, Rudolf Virchow University Hospital, Berlin		18. International Maritime Court, Hamburg
			19. Television Museum, Berlin
1988	34. Technical University Library, Berlin		20. Deutsche Bundesbank, Frankfurt/Main
	35. Multi-story parking garage, Paderborn	1990	21. Zementfabrik, Bonn
	36. New Orangery, Herten		22. Munsterlandhalle, Münster
1989	37. Leisure swimming pool, Wyk auf Föhr	1992	23. Rutgers Werke AG, Frankfurt/Main
	38. Königsgalerie, Kassel	1993	24. Railroad station, Berlin-Spandau
1990	39. VIP-Lounge, Cologne-Wahn Airport		25. Police Headquarters, Kassel
			26. Nurnberger Insurance Company
	40. Kehrwiederspitze-Sandtorhafen development, Hamburg	1994	27. Apartment and Commercial block,Kümmellstrasse/Eppendorfer Landstrasse, Hamburg
	41. Neue Strasse, Ulm		
	42. Acropolis Museum, Athens	1995	28. New Civic Center, Scharbeutz
1991	43. Altmarkt, Dresden		29. Trade Fair, Hanover, Hall 13
	44. Krefeld South		30. Bahrenfeld inner-urban area development
	45. Business Park,Münster docklands		
	46. Marina, Herne	1996	31. Gotha Main Station and station square
	47. Harbour Station Süderelbe, Hamburg		
			32. Nord LB bank building, Friedrichswall, Hanover
1992	48. Olympia 2000, bicycle and swimming hall, Berlin		33. Residential Park Elbschloss, Hamburg
	49. Noise Control Structures, Burgerfeld-Markt Schwaben		34. Housing Steinbeker Strasse, Hamburg
	50. Cologne-Ehrenfeld urban design		
1993	51. Reichstag conversion for German Bundestag, Berlin		35. Central University Library, Potsdam
	52. Building for 'Der Spiegel', Hamburg	1997	36. Fachhochschule Altonaer Strasse 25, Erfurt
	53. Festival Hall, Recklinghausen		37. Foyer refurbishment, State Opera, Hamburg
	54. Redesign Hindenburgplatz, Münster		
		1998	38. Bankerose Train Station Square, Hamburg
	55. Nürnberger Beteiligungs AG offices		39. Münchner Tor
1994	56. Erfurt-Öst, inner-urban extension		40. Technology Centre Bertrandt AG, Ehingen
	57. Max Planck Institute, Potsdam-Golm		
		1999	41. Schwäbisch Hall, Urban Design
1995	58. Redevelopment of AEG-Kanis site, Essen		42. German Embassy in Kiev
	59. 'Cultural area' of Peat Bath Spa Löbensteín		43. Museum, Schloß Homburg
	60. Fachhochschule Ingolstadt, urban design	**四等奖**	
1996	61. Spa Hotel, Hamm	1963	1. Civic Center, Kassel
	62. Haus Crange Training Hotel, Herne	1969	2. Spa Center, Westerland/Sylt
			3. Comprehensive School, Stellshoop, Hamburg
1997	63. Psychiatric Hospital, Kiel	1971	4. Civic Center, Bonn
	64. Trade Fair Essen/Gruga Park	1975	5. Extension of Town Hall, Itzehoe
	65. Dresden Central Bus Station	1979	6. Dahlem Sports Center, Free University, Berlin
	66. Bramsche Town Hall		
	67. Institute of Physics, Berlin Adlershof		7. Protestant Church Administration, Hanover
1998	68. Termina di Fusina, Venice		8. Large sports hall, Bielefeld
三等奖		1981	9. Federal Postal Department, Bonn
1965	1. Theater, Wolfsburg	1983	10. Maria-Trost Hospital , Berlin
1966	2. Sports Hall, Bottrop	1986	11. Haus der Geschichte [history museum], Bonn
1969	3. Engineering School, Buxtehude		
1970	4. Church Community Center, Stellshoop, Hamburg	1987	12. Labor Office, Flensburg
	5. School and Educational Center, Niebüll	1988	13. Schering AG offices, Berlin
		1989	14. documenta exhibition gallery, Kassel
1971	6. Development of Hamburg's inner city [west]		
			15. University Library, Kiel
1973	7. Colonia Insurance offices, Hamburg, City Nord	1993	16. Town Hall, Halle
1977	8. Post Office Savings Bank, Hamburg, City Nord	1994	17. Hypo Bank, Frankfurt/Main
			18. mdr Middle German Broadcasting House, Leipzig
1978	9. Extension Alstertal shopping mall, Hamburg		
		1995	19. Office building, An der Stadtmarine, Erfurt
	10. City Hall, Mannheim		
1979	11. Duppel-Nord Sports Center, Free University, Berlin		20. Federal Chancellery, Bonn
			21. Fachhochschule Ingolstadt
		1996	22. Ferry harbour, Mukran,Rugen
			23. Airport, Düsseldorf
			24. mdr Middle-German Broad casting Complex, Erfurt
		1997	25. Grammar-School Waltersorfer Chaussee, Berlin
		五等奖	
		1966	1. School and Sports Center, Brake
		1967	2. Open-air swimming pool, Bad Bramstedt
		1980	3. Federal Ministry of Labor and Social Affairs, Bonn
		1985	4. Library, Göttingen
		1987	5. Civic Center, Wieslock
		1989	6. Art Museum and Town Hall extension, Wolfsburg
			7. Extension Christian-Albrecht-University, Kiel
		1990	8. Zürich-Haus, Frankfurt/Main
			9. Ericusspitze, Hamburg
		1995	10. MP and Regional State Department building, Mainz
			11. Fachhochschule Ingolstadt, high-rise
		1996	12. Expo 2000, Train Station. Hanover-Laatzen
		1997	13. Elementary School Erkelenz
		荣誉奖	
		1963	1. Residenzplatz,Wurzburg
		1965	2. Löwenwall, Braunschweig
			3. Urban design, Hamburg-Niendorf
		1966	4. New Picture Gallery, Munich
			5. Urban design, Kiel inner city [special purchase]
		1967	6. Buildings for the 20th Olympic Games, Munich,project A
		1968	7. Schools Center, Weinheim
			8. Apartment blocks, An der Alster/Fontenay, Hamburg
		1969	9. Comprehensive School, Mümmelmannsberg, Hamburg
		1970	10. Schools Center, Aldeby, Flensburg
			11. Secondary School, Bargteheide1
		1971	12. Shopping mall, Hamburg-Lohbrügge
			13. Tornesch urban design ideas1
			14. Indoor swimming pool, Bad Oldesloe
		1972	15. Spa Gardens, Helgoland
		1977	16. Bauer publishing house, Hamburg
			17. Axel Springer publishing house, Hamburg [special purchase]
		1979	18. Computer Center, Deutsche Bank, Munich
			19. Civic Center, Neumunster
		1980	20. Valentinskamp, Hamburg, urban design
			21. Römerberg development, Frankfurt/Main , compulsory design
			22. Urban villas Rauchstrasse, Berlin-Tiergarten
		1981	23. Central Post Office sorting halls, Munich
		1982	24. National Library, Frankfurt/Main
			25. Savings Bank extension, Hamburg
		1983	26. Health Insurance Corp., Hamburg
		1987	27. Municipal Library, Munster - 2nd phase
			28. Pfalztheater Kaiserslautern
		1991	29. Railroad stations "Rosenstein" and "Nordbahnhof," Stuttgart
			30. State Museum of the 20th Century, Nuremberg
		1992	31. Sony, Potsdamer Platz, Berlin
		1993	32. Harness Course, Farmsen, Hamburg
			33. Festival Complex, Recklinghausen
			34. Hoffmannstrasse, Berlin-Treptow
			35. Urban design Dreissigacker-South, Meiningen

36. Theater of the City of Gütersloh
37. Kleist-Theater, Frankfurt/Oder
38. University Library and urban design ideas competition, Erfurt University
39. Redevelopment of former Domestic and Breeding Livestock Market, Lübeck
40. Museum "Alte Kraftpost", Pirmasens (1st purchase)
41. Trade Fair and Administration Center, Bremen
42. Main Train Station, Erfurt
43. Multi-Sports Hall, Leipzig
44. Prison, Gräfentonna
45. Double bridge in Fürst-Pückler-Park, Bad Muskau
46. HAB School of Architecture and Construction, Weimar
47. DVG 2000 Administration, Hanover
48. Ostra-Allee, Dresden
49. Gothaer Platz, Erfurt
50. Multy-story parking garage, Trier
51. Synagogue, Dresden
52. Main Train Station, Stuttgart 21
53. Wiso, Nuremberg
54. Potsdam Train Station Quater
55. Faculty building, University of Erlangen
56. Federal Offices of Schleswig-Holstein and Lower Saxony in Berlin
57. Munich Airport, Extension Terminal 2
58. Tromsoe Town Hall, Norway
59. Vienna Airport

完工项目

1967 · Stormann Hall, Bad Oldesloe
1969 · Max Planck Institute for Aeronomy, Lindau/Harz
       W Köhnemann House, Frankfurt/Main
1970 · Diekirch Sports Center, Luxembourg
1971 · Apartment blocks, An der Alster-Fontenay, Hamburg
1972 · Apartment block, Alstertal, Hamburg
1974 · Shell AG head offices, Hamburg
     · Airport, Berlin-Tegel
1975 · ARAL AG head offices, Bochum
     · Schools Center, Friedrichstadt
     · Airport, Berlin-Tegel, Power Station/Technical Services
     · Airport, Berlin-Tegel, Noise Control Hangar
     · Airport, Berlin-Tegel, Aircraft Maintenance Hangar
     · Airport, Berlin-Tegel, Grit Store
1976 · Regional Tax Bureau, Oldenburg
     · Sports Forum, University of Kiel
1977 · District Vocational School, Bad Oldesloe
     · Psychiatric hospital in Rickling
1978 · Urban villas, Hamburg Bau 78
     · Row houses, Hamburg Bau 78
     · Airport, Berlin-Tegel, taxi station roofing
     · Vocational Training Center G 13, Hamburg-Bergedorf
1979 · Reconstruction of the 'Fabrik', Hamburg
     · Housing Kohlhöfen, Hamburg
1980 · Hanse Viertel, Hamburg
     · European Patent and Trademark Office, Munich
     · Taima and Sulayyil, Saudi Arabia, two new desert settlements
     · Offices of the MAK, Kiel-Friedrichsort
     · Institute of Biochemistry, University of Brunswick
1981 · Renaissance-Hotel Ramada, Hamburg
     · House 'G', Hamburg-Blankenese
     · Psychiatric hospital in Rickling, semi-open ward
1982 · Otto mail-order company, head office extension, Hamburg
     · Parish House, Ritterstrasse, Stade
     · 'Black box' - Schauland sales hall, Hamburg
1983 · Regional Department of the Interior, Kiel
     · Home for the handicapped, Am Südring, Hamburg
     · Office building, Hohe Bleichen, Hamburg
     · Multi-story parking garage, Poststrasse, Hamburg
     · DAL office center, Mainz
1984 · Deutsche Lufthansa office building, Hamburg
     · Hillmann Garage, Bremen
     · Housing complex, Bad Schwartau
     · Low-Energy House, IBA Berlin
     · Urban villas, IBA Berlin
     · Sports facilities, Bad Schwartau
     · Police Station, Pankstrasse, Berlin
1985 · Plaza Hotel, Hillmannplatz, Bremen
     · Cocoloco, bar and boutique, Hanse Viertel, Hamburg
     · Psychiatric hospitals, Thetmarshof and Falkenhost in Rickling
1986 · Post Office Parking garage, Brunswick
     · Reconstruction of Michaelsen Country House as a Dolls' Museum, Hamburg
     · Trade and Vocational Training Center, Flensburg
     · Hamburg Agency in Bonn
1987 · Apartment and office building, Grindelallee 100, Hamburg

1988 · Complex Rose, hospital for rheumatic diseases, Bad Meinberg
     · Refurbishment shopping mall, Hamburger Strasse, Hamburg
     · Apartment building, Saalgasse, Frankfurt/Main
1989 · District Court, Flensburg
     · Housing Am Fischmarkt, Hamburg
     · Glass Roof, Museum of Hamburg History, Hamburg
1990 · Elbchaussee 139, Hamburg
     · Restaurant, 'Le Canard', Elbchaussee 139, Hamburg
     · Telecommunications office, Post office 1, Regional Postal Directorate, Brunswick
     · HEW Training Center, Hamburg
     · Housing and commercial building Moorbek Rondeel, Norderstedt
     · Multi-story parking garage, Hamburg-Fuhlsbüttel Airport
     · Lazarus Hospital, Berlin
     · Civic Center, Bielefeld
     · Airport, Berlin-Tegel, crew services
1991 · Airport, Stuttgart
     · Large sports hall, Flensburg
     · Café Andersen - shopping mall, Hamburger Strasse, Hamburg
     · Deutsche Lufthansa office building, Hamburg, 2nd stage
     · Subway station, Bielefeld
     · Ankara Kavaklidere Complex, Sheraton Hotel and shopping mall, Ankara
     · Saar-Galerie, Saarbrücken
     · City Center, Schenefeld
     · Apartment and commercial building, Matzen, Buchholz
     · Hillmannhaus, Bremen
     · Miro Data Systems, Brunswick
1992 · von Gerkan residence, Elbchaussee 139, Hamburg
     · Jumbo Shed, Deutsche Lufthansa Maintenance Hangar, Airport, Hamburg-Fuhlsbüttel
     · Salamander building, Berlin
     · DAL office center, Mainz, extension
     · Lazarus Hospital, Berlin, refurbishment of old hospital building
     · City Rail Station, Stuttgart Airport
1993 · Zürich-Haus, Hamburg
     · Fleetinsel - Steigenberger Hotel, Hamburg
     · Labor Office, Oldenburg
     · Telecommunications office 2, Regional Postal Directorate, Hanover
     · EAM, Kassel
     · Airport, Berlin-Tegel, parking garage P2
     · Airport, Hamburg-Fuhlsbütte
     · Collegium Augustinum, Hamburg
     · Airport, Stuttgart, 2nd stage
     · Trade and Vocational Training Center, Flensburg, 2nd stage
1994 · Hillmann-Eck, Bremen
     · Hypo-Bank, Hamburg - 'Graskeller'
     · Music and Conference Hall, Lübeck
     · Galeria, Duisburg
     · Law Courts, Brunswick
     · Deutsche Revision, Frankfurt/Main
     · Bank and commercial building, Brodschrangen/Bäckerstrasse, Hamburg
     · Apartment and commercial building Schaarmarkt, Hamburg
     · Trade and Vocational Training Center, Flensburg, 3rd stage
     · Refurbishment Law Courts, Flensburg
     · Rehabilitation clinic, Trassenheide, Usedom
     · Buildings for 'Premiere' TV channel, Studio Hamburg
1995 · German-Japanese Center, Hamburg
     · City villa Dr. Braasch, Eberswalde
     · Trade Fair, Hanover, Hall 4

337

1996
- New Trade Fair, Leipzig
- Hapag-Lloyd offices, Rosenstrasse, Hamburg
- Allee-Center, Leipzig-Grünau

1997
- Quartier 203/Atrium, Friedrichstrasse/Leipziger Strasse, Berlin
- Dresdner Bank, Pariser Platz, Berlin
- Dr. med. Manke consulting office
- Star houses, Norderstedt
- Restaurant Vau, Jägerstrasse, Berlin
- Several platform roofs for the Deutsche Bahn AG
- Housing complex, Friedrichshain, Berlin
- Commercial building, Neuer Wall 43, Hamburg
- Nordseepassage (mall), Wilhelmshaven
- Forum Köpenick, Berlin
- Town center of Schöneiche near Berlin
- Gerling Insurance, Am Löwentor, Stuttgart
- Telekom operating building, Suhl
- Bridge across the River Hörn, Kiel
- Restructuring/refurbishing Thalia-Theater, Hamburg
- Restructuring Hapag Lloyd head office, Hamburg

1998
- Lecture Theatre Centre, University of Oldenburg
- Lecture Theatre Centre, extension of Chemnitz Technical University
- Telephone offices 3 + 5, Telekom, Berlin
- Office building, Bredeney, Essen
- Residence in Jurmala, Riga
- Building for cooperative Norddeutsche Metall-Berufsgenossenschaft, Hanover
- Railroad Station, Berlin-Spandau
- Railroad bridge across the River Havel
- Connecting structures for Halls 3, 4, 5 Trade Fair, Hanover

1999
- Connecting structures for Halls 5, 6, 7 Trade Fair, Hanover,
- Calenberger Esplanade, Hannover
- Hanseatic Trade Center, Kehrwiederspitze, Hamburg
- Metropolitan Express Train, interior
- Philips Convention Stand
- Hanover Trade Fair, Hall 8/9
- Alvano House, Hamburg
- Expo 2000 - roofing over City-Railroad Stop, Hanover
- Civic Center, Weimar
- Entertainment-Center, Hamburg
- Astron Hotel, Berlin
- New supermarkets complex, Göttingen
- Office and commercial building, Friedrichstrasse 108/Johannisstrasse, Berlin
- Elbkaihaus, Hamburg

截至 1999 年的在建项目

- Foot bridges, Expo 2000
- Dar El Beida Airport, Algiers
- Noise Barrier, Station Spandau
- »Lehrter Bahnhof« - central station, Berlin
- Station 2000 - platform furnishings
- Several (ICE-)platform roofs for the Deutsche Bahn AG
- Dorotheen Blocks, parliamentarians offices, Berlin
- Ku'damm Eck Hotel, Berlin
- Labor Office Training Center of the Fed. Rep of Germany, Schwerin
- Altmarkt, Dresden, south-western side
- Stuttgart Airport, extension
- German School, Beijing
- Trade Fair, Rimini
- Trade Fair, Dusseldorf
- Station square, Koblenz
- Lenné-Passage, Frankfurt/Oder
- Railroad bridge, Lehrter Station, Berlin
- Casino, Bad Steben
- Media Center, Leipzig
- Christian Pavilion, Expo 2000
- Agency of Brandenburg and Mecklenburg-Vorpommern in Berlin
- Art Kite Museum, Detmold
- Mining Archives, Clausthal-Zellerfeld

截至 1999 年的规划阶段项目

- Shopping mall, Harburger Hof
- Development Bei St. Annen/ Holländischer Brook, Hamburg
- Residence, Dr. Manke, Uelzen
- Stuttgart Airport, terminal 3
- Conversion Main Railroad Stations Kiel, Mainz, Lübeck
- Central Bus Station, Wilhelmshaven
- Deutrichs Hof, Leipzig
- Agency of Sachsen-Anhalt in Berlin
- German Trade Center, Bucharest
- International Horticulture Exhibition, Rostock
- Kroepcke Center, Hanover
- Apartment House, Riga
- Office and Commercial Building, Riga
- Berlin Olympic Stadium, conversion
- Berlin-Brandenburg International Airport
- Nanning Convention & Exhibition Center, China
- Ancona Airport, Italy

获奖称号

- German Architecture Award 1977, commendation:
  Power station/technical services building Airport Berlin-Tegel

- Gold medal of the Federal Competition Industry in Urban Design 1978:
  Airport, Berlin-Tegel

- 'Exemplary Buildings', distinction:
  Housing Kohlhöfen, Hamburg

- 'Exemplary Buildings' 1978, distinction:
  Vocational training center G 13, Hamburg-Bergedorf

- Building of the Year 1979 (AIV):
  Vocational training center G 13, Hamburg-Bergedorf

- BDA Award Schleswig-Holstein 1979:
  Sports forum, University of Kiel

- Architecture Award Concrete 1979, commendation:
  Sports forum, University of Kiel

- 'Exemplary Buildings', distinction: Urban villas, Hamburg, Bau 78

- Poroton Architects' Competition, 1st prize:
  Urban villas, Hamburg, Bau 78

- 'Exemplary Buildings', distinction:
  Row houses, Hamburg, Bau 78

- Poroton Architects' Competition, special prize:
  Row houses, Hamburg, Bau 78

- BDA Award Lower Saxony 1980:
  Max Planck Institute, Lindau/Harz

- BDA Award Bavaria 1981, commendation:
  European Patent and Trademark Office, Munich

- International Color Design Award 1980/81, commendation:
  Airport, Berlin-Tegel

- Building of the Year 1981 (AIV):
  Hanse Viertel, Hamburg

- Building of the Year 1983 (AIV):
  Multi-story parking garage, Hamburg

- Mies van der Rohe Award 1984:
  Hanse Viertel, Hamburg

- North-German Timber Construction Award 1984:
  House 'G', Hamburg-Blankenese

- BDA Award Schleswig-Holstein 1985:
  Marktarkaden, Bad Schwartau

- BDA Award Schleswig-Holstein 1985:
  Regional Department of the Interior, Kiel

- 'Exemplary Buildings' 1989, commendation:
  Grindelallee 100, Hamburg

Mies van der Rohe Award 1990:
Glass roof, Museum of Hamburg History,
Hamburg

Building of the Year 1990 (AIV):
Elbchaussee 139, Hamburg

BDA Award Bremen 1990:
Hillmann-Garage, Bremen

BDA Award Lower Saxony 1991:
Regional Postal Directorate, Brunswick

Deutscher Natursteinpreis 1991:
Regional Postal Directorate, Brunswick

German Architecture Award 1991:
Multi-story parking garage,
Hamburg-Fuhlsbüttel

BDA Award
North Rhine-Westphalia 1992,
'Bauen für die öffentliche Hand'
[public buildings],
commendation:
Civic Center, Bielefeld

BDA Award
North Rhine-Westphalia 1992,
'Bauen für die öffentliche Hand'
[public buildings],
commendation:
Subway station, Bielefeld

BDA Award Berlin 1992,
commendation:
Salamander building, Berlin

German Steel Construction Award 1992:
Airport, Stuttgart

Westhyp Architecture Award 1992,
commendation: HEW Training Center,
Hamburg

Deutscher Natursteinpreis 1993,
commendation:
Airport, Stuttgart

Deutscher Verzinkerpreis 1993,
commendation:
Zürich-Haus, Hamburg

Deutscher Verzinkerpreis 1993,
commendation:
'Le Canard' bridge, Hamburg

Building of the Year 1993 (AIV):
Jumbo shed, Hamburg

Deutscher Verzinkerpreis 1993:
Multi-story parking garage,
Hamburg-Fuhlsbüttel

Balthasar Neumann Award 1993:
Airport, Hamburg-Fuhlsbüttel

Building of the Year 1994 (AIV):
Airport, Hamburg-Fuhlsbüttel

Prix d'Excellence 1994 -
Finaliste Catégorie
Immobilier d'Entreprise:
Jumbo shed, Hamburg

BDA Award Lower Saxony 1994,
commendation:
Labor Office, Oldenburg

Peter Joseph Krahe Award 1994:
Regional Postal Directorate, Brunswick

Peter Joseph Krahe Award 1994:
Miro Data Systems, Brunswick

- BDA Award Lower Saxony 1994:
  Miro Data Systems, Brunswick

- Constructec Award 1994,
  commendation:
  Miro Data Systems, Brunswick

- Deutscher Natursteinpreis 1995,
  commendation:
  Law Court, Brunswick

- Oldenburger Stadtbildpreis 1995,
  commendation:
  Labor Office, Oldenburg

- German Architecture Award 1995,
  commendation:
  Airport, Hamburg-Fuhlsbüttel

- USITT 1996, Honor Award:
  Music and Conference Hall, Lubeck

- German Steel Construction Award 1996,
  commendation:
  Airport, Hamburg-Fuhlsbüttel

- Saxony State Award for Architecture and
  Construction 1996:
  New Trade Fair, Leipzig

- Westhyp Architectural Award 1996,
  commendation:
  Bank and Commercial building,
  Brodschrangen, Hamburg

- BDA Award Hamburg 1996:
  Jumbo shed, Hamburg

- BDA Award Hamburg 1996,
  commendation:
  Airport, Hamburg-Fuhlsbüttel

- BDA Award Lower Rhine 1996,
  commendation:
  Galeria, Duisburg

- Brunel Awards 1996:
  Platform roof prototype for
  Deutsche Bahn AG

- Deutscher Natursteinpreis 1996,
  commendation:
  Hapag-Lloyd offices, Hamburg

- Design Innovations 1997,
  Red Point for high design quality:
  glass-door band, type 16058

- Design Innovations 1997,
  Red Point for high design quality:
  Platform roof prototype for
  Deutsche Bahn AG

- Deutscher Verzinkerpreis 1997:
  New Trade Fair, Leipzig

- German Architecture Award 1997,
  distinction:
  New Trade Fair, Leipzig

- Martin-Elsaesser-Plakette 1998,
  commandation: C&L Deutsche Revision

- Bathasar-Neumann Award 1998:
  Trade Fair, Hanover, Hall 4

- German Steel Construction Award 1998,
  commendation: New Trade Fair, Leipzig

- The 1998 Dupont Benedictus Awards:
  Dresdner Bank, Pariser Platz, Berlin

- The 1998 Dupont Benedictus Awards:
  Nordseepassage, Wilhelmshaven

- Deutscher Natursteinpreis 1999,
  commendation: Railroad Station, Berlin-
  Spandau

- Marble Architectural Awards 1999,
  special mention: New Trade Fair, Leipzig

- Deutscher Verzinkerpreis 1999:
  Railroad Station, Berlin-Spandau

展馆建筑

- Architecture of von Gerkan,
  Marg and Partners, BDA, Dresden 1989

- Architecture of von Gerkan, Marg and
  Partners, Weimar Bauhaus, BDA 1990

- Idea and Model -
  30 Years of Architectural Models of
  von Gerkan, Marg and Partners,
  Architecture Workshop, Hamburg 1994

- Under Big Roofs, wide-tensed projects of
  von Gerkan, Marg and Partners, Berlinische
  Gallery, Berlin 1995

- Infobox Berlin:
  Lehrter Bahnhof - new central train station
  in Berlin, Berlin 1996/1997

- Building for Air Travel -
  Architecture and Design for Commercial
  Aviation: Airport Tegel, Stuttgart, Hamburg,
  Jumbo Shed,
  The Art Institute of Chicago 1996/1997
  Museum of Flight, Seattle 1997
  San Francisco International Airport 1997
  Rhein-Main International Airport,
  Frankfurt/Main 1997
  Tempelhof Airport, Berlin 1997
  Hamburg Airport 1997

- Sensing the Future -
  The Architect as Seismograph:
  New Trade Fair, Leipzig,
  Architecture Biennial, Venice 1996

- Renaissance of Railway Stations.
  The City in the 21st Century,
  22 buildings and projects of von Gerkan,
  Marg and Partners, Architecture Biennial,
  Venice 1996
  Former Dresdner Train Station in Berlin 1997
  Main Train Station, Stuttgart 1997
  Deichtor halls, Hamburg 1997
  Former Deutsche Bahn cargo hall,
  Munich 1998
  Colosseum, Essen 1999

- Building According to Nature -
  The Heirs of Palladio in Northern Europe:
  New Trade Fair, Leipzig, Museum of Hamburg
  History, Hamburg 1997

- Berlin Scrapes the Sky, DAZ, Berlin 1997

- Airports - Vision and Tradition -
  Designs for an Berlin-Brandenburg
  International Airport:
  Airport Berlin-Tegel, Stuttgart, Hamburg and
  21 student projects under Prof. M. v. Gerkan,
  IDZ, Berlin-Tempelhof Airport, Berlin 1997

- Macht und Monument: Lehrter Bahnhof,
  German Architecture Museum, Frankfurt
  1998

- The Architect as Designer: The VAU chair,
  Shoe-cleaning step stool,
  Museum of Arts and Crafts, Berlin 1998

- 2002 800 Years Riga, "A House for the Music", diploma designs and Music and Conference hall in Lübeck, Rigas Galerija, Riga 1998

- Bucharest 2000, German-Romanian Architectural Workshop, a project of the Jürgen-Ponto-Foundation Supervision: Meinhard von Gerkan Sala Dalles Artexpo, Bucharest 1998

- Architecture of Contemplation, Gallery Aedes East, Berlin 1998

- 100 Years Art on the Move: Conversion of the Reichstag building, Lehrter Bahnhof, Berlin Gallery Bonn, Bonn 1998

- In/From China: Art and Architecture German School of Beijing, Asian Fine Arts Factory, Berlin 1999

- "The Christian Pavilion" on the World Exhibition EXPO 2000 Evangelic Academy, Hanover 1999

- von Gerkan, Marg and Partners · Building for the Public, Yan-Huang Art Museum, Beijing, China 1999

相关著作

- Meinhard von Gerkan Architektur 1966 – 1978 von Gerkan, Marg und Partner Stuttgart: Karl Krämer Verlag 1978 ISBN 3-7828-1438-X

- Meinhard von Gerkan Die Verantwortung des Architekten - Bedingungen für die gebaute Umwelt Stuttgart: Deutsche Verlags-Anstalt 1982 ISBN 3-421-02584-3

- Meinhard von Gerkan Architektur 1978 – 1983 von Gerkan, Marg und Partner Stuttgart: Deutsche Verlags-Anstalt 1983 ISBN 3-421-02597-5

- Meinhard von Gerkan Alltagsarchitektur Gestalt und Ungestalt Wiesbaden/Berlin: Bauerverlag 1987 ISBN 3-7625-2449-1

- Meinhard von Gerkan Architektur 1983 – 1988 von Gerkan, Marg und Partner Stuttgart: Deutsche Verlags-Anstalt 1988 ISBN 3-421-02893-1

- Meinhard von Gerkan Jürgen-Ponto-Stiftung West-Östlicher Architektenworkshop in Dresden 13. – 20. Juli 1990 Hamburg: Christians Verlag 1990 ISBN 3-7672-1121-1

- Meinhard von Gerkan Architektur 1988 – 1991 von Gerkan, Marg und Partner Stuttgart: Deutsche Verlags-Anstalt 1992 ISBN 3-421-03021-9

- Volkwin Marg Architektur in Hamburg seit 1900 Hamburg: Junius Verlag GmbH 1993 ISBN 3-88506-206-2

- Meinhard von Gerkan von Gerkan, Marg and Partners London: Academy Editions Berlin: Ernst & Sohn 1993 ISBN 1-85490-166-4

- Meinhard von Gerkan Idea and Model Idee und Modell 30 years of architectural models 30 Jahre Architekturmodelle von Gerkan, Marg und Partner Meinhard v. Gerkan unter Mitarbeit von Jan Esche und Bernd Pastuschka Berlin: Ernst & Sohn 1994 ISBN 3-433-02482-0

- Ideen, Entwürfe, Modelle Meinhard von Gerkan,gmp Die Musik- und Kongresshalle (MUK) (Ausstellungskatalog) Hg. v.Overbeck-Gesellschaft Verein von Kunstfreunden e.V.,Lübeck Lübeck 1994

- Meinhard von Gerkan Architektur im Dialog Texte zur Architekturpraxis Berlin: Ernst & Sohn 1994 ISBN 3-433-02881-8

- von Gerkan, Marg und Partner Unter großen Dächern (Ausstellungskatalog) Hg. v. Klaus-Dieter Weiß in Kooperation mit Berlinische Galerie, Landesmuseum für Moderne Kunst, Photographie und Architektur Gütersloh,Braunschweig/Wiesbaden: Bertelsmann Fachzeitschriften GmbH und Friedr.Vieweg & Sohn Verlagsgesellschaft mbH 1995 ISBN 3-528-08194-5

- C & L Deutsche Revision Bürogebäude Heddernheim Hg. v. C & L Deutsche Revision Frankfurt 1995

- Meinhard von Gerkan Architektur 1988 – 1991 von Gerkan, Marg und Partner 2., veränderte Auflage Basel, Boston, Berlin: Birkhäuser 1995 ISBN 3-7643-5221-3

- Meinhard von Gerkan Culture Bridge Deutsch-Polnischer Ideenwettbewerb "Kulturbrücke Görlitz/Zgorzelec" Berlin: Vice Versa Verlag 1995

- architypus special von Gerkan, Marg und Partner Unter großen Dächern. Hg. v. Klaus-Dieter Weiß Gütersloh, Braunschweig/Wiesbaden: Bertelsmann Fachzeitschriften GmbH und Friedr. Vieweg & Sohn Verlagsgesellschaft mbH 1995 ISBN 3-528-08113-9

- Meinhard von Gerkan Architektur 1991 – 1995 von Gerkan, Marg und Partner Basel, Boston, Berlin: Birkhäuser 1995 ISBN 3-7643-5222-1 ISBN 0-8176-5222-1

- Volkwin Marg Pilot-Projekt Aufbau Ost – Neue Messe Leipzig, Planung + Bau 1992-1995 Basel, Boston, Berlin: Birkhäuser 1996 ISBN 3-7643-5409-7

- Renaissance of Railway Stations. The City in the 21st Century (Ausstellungskatalog) Hg. v. Bund Deutscher Architekten BDA, Deutsche Bahn AG, Förderverein Deutsches Architekturzentrum DAZ in cooperation with Meinhard v. Gerkan Wiesbaden:Vieweg 1996

- Meinhard v. Gerkan von Gerkan, Marg und Partner Architecture for Transportation, Architektur für den Verkehr Basel, Boston, Berlin: Birkhäuser 1997 ISBN 3-7643-5611-1 ISBN 0-8176-5611-1

- Volkwin Marg Neue Messe Leipzig, New Trade Fair Leipzig Basel, Boston, Berlin: Birkhäuser 1997 ISBN 3-7643-5429-1 ISBN 0-8176-5429-1

- John Zukowsky The Architecture of von Gerkan, Marg + Partners Prestel Verlag, München, New York 1997 ISBN 3-7913-1861-6

- Meinhard von Gerkan Architecture 1995 – 1997 von Gerkan, Marg und Partner Basel, Boston, Berlin: Birkhäuser 1998 ISBN 3-7643-5844-0

- Meinhard von Gerkan Möbel Furniture von Gerkan, Marg und Partner Verlag Gert Hatje, Stuttgart 1998 ISBN 3-7757-0766-2

- Von Gerkan, Marg and Partners · Building for the Public (Ausstellungskatalog) Hg. v. Yan-Huang Art Museum, Beijing 1999

- Volkwin Marg Hall 8/9, von Gerkan, Marg + Partners Munich, London, New York 2000 ISBN 3-7913-2136-6

1965-1999年竞赛作品一览

Abbreviations:
D for: design; P for: partner;
PL for: project leader; C for: co-worker;
Co for: in cooperation with

1965 年以前

1. District Administration, Niebüll
   competition, 2nd prize

2. Civic Center, Kassel
   competition, 4th prize

3. Residenzplatz, Würzburg
   competition, award

4. Löwenwall, Brunswick
   competition, award

5. Airport, Hanover-Langenhagen
   M. v. Gerkan, diploma project

6. Indoor swimming pool, Brunswick-Gliesmarode
   competition, 1st prize

7. Sports- and Conference Center, Hamburg
   competition, 1st prize

8. Jungfernstieg, Hamburg
   competition, 2nd prize

9. Theater, Wolfsburg
   competition, 3rd prize

10. Airport, Berlin-Tegel
    competition, 1st prize
    completed 1975
    D: M. v. Gerkan, K. Nickels
    P: K. Staratzke, K. Brauer, R. Niedballa
    C: W. Hertel, H. Herzlieb, W. Hönnicke, M. Illig, D. Perisic, P. Römer, G. Seule, H. Pitz, W. Zimmer, H.-J. Roeske

11. Airport, Berlin Tegel, Noise Control Hangar
    completed 1975
    D: M. v. Gerkan
    P: K. Brauer

12. Airport, Berlin-Tegel, Power Station/Technical Services
    completed 1975
    D: M. v. Gerkan
    P: K. Staratzke
    C: W. Hönnicke, L. Gerhardt, M. Auder, C. Grzimek, R. Henning

13. Airport, Berlin-Tegel, bridges
    completed 1975
    D: M. v. Gerkan
    P: K. Staratzke

14. Airport, Berlin-Tegel, Freight Center
    completed 1972
    D: M. v. Gerkan
    P: K. Staratzke
    C: W. Hönnicke

15. Airport, Berlin-Tegel, Grit Store
    completed 1972
    D: M. v. Gerkan, K. Staratzke
    P: K. Staratzke

16. Airport, Berlin-Tegel, apron areas
    completed 1973
    D: M. v. Gerkan
    P: R. Niedballa
    C: P. Römer

17. Urban design, Kiel inner city
    competition, award

18. Indoor/open-air swimming pool, Sozialdemokratische Partei Deutschlands
    competition, 1st prize

19. Stormarn Hall, Bad Oldesloe
    competition, 1st prize
    completed 1967
    D: V. Marg with H. Schmedje, K. Nickels

20. Diekirch Sports Center, Luxembourg
    competition, 1st prize
    completed 1970 and 1975 (2nd stage)
    D: M. v. Gerkan
    C: C. Brockstedt, C. Claudius, S. Müllerstedt
    Co: R. Störmer

21. Tax Bureau, Oldenburg
    competition, 1st prize
    completed 1976
    D: M. v. Gerkan
    C: C. Mrozek
    Co: D. Patschan

22. Urban design, Hamburg-Niendorf
    competition, award

1966

23. School and Sports Center, Brake
    competition, 5th prize

24. Max Planck Institute for Aeronomy in Lindau/Harz
    competition, 1st prize
    completed 1969
    D: V. Marg
    C: C. Claudius, G. Fleher
    Co: R. Störmer

25. District indoor swimming pool, Cologne
    competition, 1st prize

26. Sports Hall, Bottrop
    competition, 3rd prize

27. New Picture Gallery, Munich
    competition, award

28. Sports Forum, University of Kiel
    competition, 1st prize
    completed 1976
    D: V. Marg with K. Nickels
    C: K. Kurzweg, G. Welm, V. Rudolph, P. Frohne

1967

29. Open-air swimming pool, Bad Bramstedt
    competition, 5th prize

30. Buildings for the 20th Olympic Games, Munich
    competition, project A: award project B: 2nd prize

31. University of Bremen
    competition

1968

32. Schools Center, Weinheim
    competition, award

33. Airport, Hamburg-Kaltenkirchen
    consultancy project

34. Administrative offices, Hamburgische Landesbank, Hamburg
    competition entries, 1st place

35. Adolfinum Secondary School, Bückeburg
    competition

36. Church Center, Hamburg-Ohlsdorf
    competition, 2nd prize

37. Apartment blocks, An der Alster – Alsterufer/Fontenay, Hamburg
    competition, award
    completed 1971
    D: V. Marg
    C: C. Mrozek, P. Fischer

38. W. Kohnemann House, Hamburg
    completed 1969
    D: V. Marg

39. Parish Hall and Church, Osdorfer Born
    competition, 2nd prize

1969

40. Shopping mall, Hamburg-Altona
    competition, award

41. Dr. Hess residence, Hamburg-Nienstedten

42. Engineering School, Buxtehude
    competition, 3rd prize

43. Spa Center, Westerland/Sylt
    competition, 4th prize

44. Comprehensive School, Steilshoop
    competition, 4th prize

45. Comprehensive School, Mümmelmannsberg, Hamburg
    competition, award

46. Church Community Center, Hamburg-Bergedorf
    competition, second round

1970

47. Shell AG head offices, Hamburg
    competition, 1st prize
    completed 19./4
    D: V. Marg
    P: E. Wiehe
    PL: B. Albers
    C: J. Lupp, H.-P. Harm, K. Maass, U. Rückel, H. Stetten

48. Mobil Oil AG offices, Hamburg
    competition

49. District Vocational School, Bad Oldesloe
    competition, 1st prize
    completed 1977
    D: V. Marg
    P: E. Wiehe
    U: F. Ferdinand, E. Schäfer, D. Wingsch

50. Schools Center, Heide-Ost
    competition, 2nd prize

51. Regional Postal Directorate, Bremen
    competition, 2nd prize

52. Church Community Center, Steilshoop, Hamburg
    competition, 3rd prize

53. School and Educational Center, Niebüll
    competition, 3rd prize

54. Redesign of Gerhard-Hauptmann-Platz, Hamburg
    consultancy project

55. Schools Center, Adelby, Flensburg
    competition, award

56. Secondary School, Bargteheide
    competition, award

57. Program for Berlin public swimming pools
    expertise

1971

58. Airport, Berlin-Tegel, Aircraft Maintenance Hangar
    completed 1975
    D: M. v. Gerkan
    P: K. Staratzke, K. Brauer, R. Niedballa
    C: R. Henning

59. Shopping mall, Hamburg-Lohbrügge
    competition, award

60. Hamburg-Poppenbüttel
    urban design expertise

61. Development of Hamburg's inner city (west)
    competition, 3rd prize

62. Tornesch
    urban design ideas competition, award

63. Multi-purpose building III, University of Hamburg
    competition, 1st prize

64. Federal Chancellery, Bonn
    competition, 4th prize

65. Cultural Center, Munich-Gasteig
    competition

66. Regional Tax Bureau, Hamburg
    competition, 2nd prize

67. European Patent and Trademark Office, Munich
    competition, 1st prize
    completed 1980
    D: V. Marg
    P: A. Sack
    C: H. Tomhave, H. Müller-Röwekamp, Korus, F. Kessler, K. Bachmann, H.-J. Roeske, K. Springhorn

68. Indoor swimming pool, Bad Oldesloe
    competition, award

69. Sports Forum, University of Bremen
    competition, 2nd prize

70. Apartment blocks, Gellertstrasse/Bellevue, Hamburg
    competition, 1st prize

71. Regional Administration offices, Lüneburg
    competition, 2nd prize

72. Housing Neuwiedenthal-Nord, Hamburg-Harburg
    urban design expertise

73. Shopping mall, Alstertal, Hamburg
    competition, 1st prize

74. GPD-Bremen
    competition 2nd prize

1972

75 Apartment block next to shopping mall, Alstertal, Hamburg
consultancy project
completed 1972
D: V. Marg
C: U. Ferdinand

76 Spa Gardens, Helgoland
competition, award

77 Regional Postal Directorate, Hamburg
competition, 2nd prize

78 Schools Center, Friedrichstadt
competition, 1st prize
completed 1975
D: V. Marg
C: R. Wilkens, H. Wolf, A. Marg, C. Mrozeck, U. Rückel

79 ARAL AG head offices, Bochum
competition, 1st prize
completed 1975
D: M. v. Gerkan
P: M. Sack, R. Niedballa
C: B. Gronemeyer, J. Busack, H. Stetten

80 Landscaped indoor swimming pool, Rebstockpark, Frankfurt/Main
competition

1973

81 Private residence for B., Hamburg-Reinbek
design

82 Colonia Insurance AG offices, Hamburg
competition, 3rd prize

83 Wilhelm-Hack-Museum, Ludwigshafen
competition

84 Tegeler Hafen Apartment blocks, Berlin
competition

85 Psychiatric hospital in Rickling
completed 1977
D: V. Marg, K. Staratzke
C: C. Mrozeck, U. Rückel, D. Wingsch

1974

86 Vocational Training Center G 13, Hamburg-Bergedorf
competition, 1st prize
completed 1978
D: M. v. Gerkan
PL: B. Albers
C: H.-E. Bock, W. Schäfer, J. Busack, M. Stroh, M. Ebeling, G. Göb, K. Maass

87 Hanse Viertel, Hamburg
completed 1980
D: V. Marg
P: K. Staratzke
C: B. Albers, R. Born, A. Buchholz-Berger, O. Dorn, H. - J. Dörr, W. Edler, M. Eggers, U. Ferdinand, R. Henning, B. Gronemeyer, J. Krautberger, A. Lucks, K. Maass, H. Müller-Röwekamp, D. Pensic, R. Seifert, P. Sembritzki

88 Cocoloco, bar and boutique, Hanse Viertel, Hamburg
completed 1985
D: K. Staratzke, O. Dorn

89 Provinzial Insurance Company offices, Kiel
consultancy project, 1st place

90 Satellite city, Billwerder-Allermöhe, Hamburg
consultancy project

91 Airport Munich II
project 1 and 2: 1974, commissioned expertise

1975

92 Airport Munich II
competition, 1st place

93 Housing Billwerder-Allermöhe, Hamburg
competition, 2nd prize

94 Regional Department of the Interior, Kiel
competition, 2nd prize
completed 1983
D: V. Marg
PL: B. Albers
C: G. Göb, W. Tegge, D. Winter

95 Town Hall of Itzehoe, extension
competition, 4th prize

96 Housing Kohlhöfen, Hamburg
completed 1979
D: V. Marg
C: J. Werner, G. Werner, H. Huussmann, B. Albers, B. Gronemeyer

97 Housing Oevelgönne, Hamburg
design

98 Regional Government offices, Brunswick
competition

99 Deutscher Ring, Hamburg
competition, 1st prize

100 Federal Department of Health, Berlin-Marienfelde
competition

101 Redevelopment Große Bleichen, Hamburg

1976

102 Administrative offices, operating area of Airport Munich II
competition, 2nd prize

103 Faculty of Inner Security, Riyadh
consultancy project

104 Airport, Moscow
competition, 1st prize

105 Dar El Beida Airport, Algiers passenger and freight terminals
competition, version A: 1st prize

106 Bauer publishing house, Hamburg
competition, award

107 District Administration, Recklinghausen
competition, 1st prize

108 Design of a power station
study

109 Post Office Savings Bank, Hamburg
competition, 3rd prize

110 Urban design, Universität Ost, Bremen
competition, 2nd prize

111 Training Center of the Federal Fiscal Administration, Münster
competition, second round

112 Holstentorplatz, Lübeck
competition

113 Communications Center, Wiesbaden
competition

1977

114 Airport, Berlin-Tegel, taxi station roofing
completed 1978
D: M. v. Gerkan
P: K. Brauer
C: M. Auder, P. Römer

115 Airport, Berlin-Tegel, crew services
completed 1978
D: M. v. Gerkan
P: R. Niedballa
C: P. Römer

116 Offices of the MAK, Kiel-Friedrichsort
consultancy project, 1st place
completed 1980
D: M. v. Gerkan
P: K. Staratzke
Co: Brockstedt + Discher

117 Parish House, Ritterstrasse, Stade
competition, 1st prize
completed 1982
D: V. Marg
P: K. Staratzke
C: E. Hamer, A. Marg, U. Rückel

118 Police Station, Panckstrasse, Berlin
competition, 1st place
completed 1984
D: M. v. Gerkan
C: M. Auder, P. Römer

119 Reconstruction of the 'Fabrik', Hamburg
completed 1979
D: V. Marg
C: J. Busack, M. Ebeling, G. Göb, C. Mrozeck, G. Sievers

120 Otto mail-order company, head office extension, Hamburg
competition, 1st prize
completed 1982
D: V. Marg
C: M. Ebeling, B. Gronemeyer, J. Kleiberg, R. Seifert, J. Sefl, C. Timm-Schwarz

121 Hyatt Hotel, Abu Dhabi

122 Axel Springer publishing house
competition, special award

123 Psychiatric Hospital in Rickling, semi-open award
completed 1981
D: V. Marg, K. Staratzke
C: K. Ehlert, E. Hamer, R. Henning, C. Mrozeck, C. Timm-Schwarz

124 Psychiatric Hospital in Rickling, Thetmarshof and Falkenhorst
completed 1985
D: V. Marg, K. Staratzke
C: K. Ehlert, E. Hamer, R. Henning

125 Row houses, Hamburg Bau 78
competition, 2nd prize
completed 1978
D: M. v. Gerkan
P: K. Staratzke

126 Urban villas, Hamburg Bau 78
competition, 2nd prize
completed 1978
D: M. v. Gerkan
P: H.-E. Bock

127 House 'G', Hamburg-Blankenese
completed 1981
D: M. v. Gerkan
C: M. Ebeling, U. Rückel, M. Stroh

128 Taima, Saudi Arabia, new desert settlement
completed 1980
D: M. v. Gerkan
P: K. Brauer, K. Staratzke
C: A. Buchholz-Berger, W. Haux

129 Sulayyil, Saudi Arabia, new desert settlement
completed 1980
D: M. v. Gerkan
P: K. Brauer, K. Staratzke
C: A. Buchholz-Berger, W. Haux

130 Sports and leisure complex, Berlin
study

131 Development project, Uhlandstrasse, Berlin
competition

1978

132 Alstertal shopping mall, extension
competition, 3rd prize

133 Hotel, Augsburger Platz, Berlin
competition

134 Restaurant on former garbage depot, Berlin-Lübars
competition

135 Federal German Ministry of Transportation, Bonn
competition, 1st prize

136 Regional Ministry for Social Affairs and Nutrition, Stuttgart, Wulle area
consultancy project

137 Faculty of Chemistry, University of Brunswick
competition, 1st place

138 Institute of Biochemistry, University of Brunswick
consultancy project, 1st place
completed 1980
D: M. v. Gerkan with H.-E. Bock, M. Stanek

139 City Hall, Mannheim
competition, 3rd prize

140 Superior Court of Justice for Berlin
competition

41　Multi-story parking garage,
　　Poststrasse, Hamburg
　　completed 1983
　　D:　M. v. Marg
　　P:　K. Staratzke
　　C:　R. Born, R. Henning, R. Seifert,
　　　　P. Sembritzki

42　Joachimsthaler Platz, Berlin
　　'Urban pavilion'
　　competition

43　Joachimsthaler Platz, Berlin
　　'Kunstwäldchen'
　　competition
　　recommended for revision

44　Joachimsthaler Platz, Berlin
　　'Light columns'
　　adopted solution

45　Pahlavi National Library, Tehran
　　international competition, 1st prize

46　Central Library, main dining hall and
　　sports facilities,
　　University of Oldenburg
　　special competition procedure

1979

47　6 urban villas, IBA Berlin
　　competition, selected
　　completed 1984
　　D:　M. v. Gerkan
　　C:　M. Auder, P. Römer

48　Museum, Aachen
　　competition

49　Home for the handicapped,
　　Am Südring, Hamburg
　　completed 1983
　　D:　V. Marg
　　P:　K. Brauer
　　C:　F. Brandt, M. Mews

50　Islamic Cultural Center, Madrid
　　competition

51　Düppel-Nord Sports Center,
　　Free University, Berlin
　　competition, 3rd prize

52　Dahlem Sports Center,
　　Free University, Berlin
　　competition, 4th prize

53　Protestant Church Administration,
　　Hanover
　　competition, 4th prize

54　Landeszentralbank Hessen,
　　Frankfurt/Main
　　competition

55　Valentinskamp, Hamburg
　　urban design, competition, award

56　Indoor swimming pool
　　Mannheim-Herzogenried
　　competition, 1st prize

57　Large sports hall, Bielefeld
　　competition, 4th prize

58　Görlitzer Bad, Berlin-Kreuzberg
　　competition

59　Indoor/outdoor swimming
　　pools, Berlin-Spandau
　　competition, 1st prize

160　Computer Center, Deutsche Bank,
　　　Hamburg
　　　competition, award

161　Civic Center, Neumünster
　　　competition, award

162　Parliament of North
　　　Rhine-Westphalia, Düsseldorf
　　　competition

163　Vereins- und Westbank, branch
　　　Ost-West-Strasse, Hamburg
　　　competition, 1st place

164　Emsland District Administration,
　　　Meppen
　　　competition, 1st prize

1980

165　Gasworks, Munich
　　　competition, 2nd prize

166　Town Hall, am Pferdemarkt,
　　　Oldenburg
　　　competition, 2nd prize

167　Office building, Hohe Bleichen,
　　　Hamburg
　　　completed 1983
　　　D:　V. Marg, K. Staratzke
　　　P:　K. Staratzke
　　　C:　E. Braunsburger, L. Flores,
　　　　　W. Haux, B. Gronemeyer,
　　　　　K. Maass, A. Wolter, A. Wriedt

168　Federal Ministry of Labor and
　　　Social Affairs, Bonn
　　　competition, 5th place

169　Administrative offices for
　　　Volkswagen, Wolfsburg
　　　competition, 2nd prize

170　Renaissance-Hotel Ramada,
　　　Hamburg
　　　completed 1981
　　　D:　V. Marg
　　　P:　K. Staratzke
　　　C:　A. Buchholz-Berger, O. Dorn,
　　　　　A. Böke, J. Krugmann,
　　　　　H. Ladewig, R. Preuss, B. Sinnwell

171　Colonia Insurance head office,
　　　Cologne-Holweide
　　　competition

172　Urban villas, Rauchstrasse,
　　　Berlin-Tiergarten
　　　competition, award

173　Civic Center, Bielefeld
　　　competition, 1st prize
　　　completed 1990
　　　D:　M. v. Gerkan
　　　PL:　M. Zimmermann
　　　C:　M. Ebeling, P. Kropp, S. Rimpf,
　　　　　T. Rinne, P. Sembritzki
　　　AOS: D. Tholotowsky, H. Schröder

174　University of Bremerhaven
　　　competition, 3rd prize

175　Academy of Fine Arts, Hamburg
　　　competition, 1st prize

176　Residential project for the
　　　Friedrichstadt neighborhood, Berlin
　　　competition

177　Trade and Vocational Training
　　　Center, Flensburg
　　　competition, 1st prize
　　　completed: 1st construction stage
　　　1986, 2nd stage 1993, 3rd stage 1994
　　　D:　M. v. Gerkan
　　　P:　K. Staratzke
　　　C:　H.-E. Bock, M. Stanek, K. Krause

178　Lazarus Hospital, Berlin
　　　competition, 1st prize
　　　completed 1990
　　　D:　M. v. Gerkan
　　　PL:　P. Römer
　　　C:　J. Zais

179　Lazarus Hospital, Berlin
　　　Refurbishment of old hospital
　　　building,
　　　completed 1992
　　　D:　M. v. Gerkan
　　　C:　P. Römer, J. Zais

180　Airport, Stuttgart
　　　competition, 1st prize
　　　completed 1990
　　　D:　M. v. Gerkan, K. Brauer
　　　P:　K. Staratzke
　　　C:　A. Buchholz-Berger, M. Dittmer,
　　　　　O. Dorn, M. Ebeling, E. Grimmer,
　　　　　G. Hagemeister, R. Henning,
　　　　　B. Kiel, A. Lucks, M. Mews,
　　　　　H.-H Möller, D. Perisic,
　　　　　K.-H. Petersen, U. Pörksen,
　　　　　S. Rimpf, H. Thimian,
　　　　　C. Timm-Schwarz, T. Tran-Viet,
　　　　　H. Ueda

181　Römerberg development,
　　　Frankfurt/Main
　　　competition, free design: 1st prize;
　　　compulsory design: award

182　Fleetinsel development, Hamburg
　　　competition, 1st prize

183　Fleetinsel – Steigenberger Hotel,
　　　Hamburg
　　　competition, 1st prize
　　　completed 1993
　　　D:　V. Marg with W. Haux
　　　C:　A. Böke, J. Krugmann,
　　　　　H. Ladewig, R. Preuss, B. Sinnwell

184　Large sports hall, Flensburg
　　　competition, 1st prize
　　　completed 1991
　　　D:　M. v. Gerkan with M. Stanek
　　　P:　K. Staratzke
　　　C:　K. Krause

185　Deutsche Lufthansa office building,
　　　Hamburg
　　　competition, 1st prize
　　　1st stage 1984; 2nd stage 1991
　　　D:　M. v. Gerkan, K. Brauer
　　　P:　K. Staratzke
　　　C:　M. Ebeling, B. Gronemeyer,
　　　　　K. Maass, M. Mews, H.-H. Möller

186　Indoor leisure swimming pool, Kiel
　　　competition

187　Max Planck Institute for Quantum
　　　Optics, Munich
　　　competition, 2nd prize

188　Sports facilities, Bad Schwartau
　　　completed 1984
　　　D:　M. v. Gerkan, K. Brauer
　　　C:　M. Ebeling, W. Haux, H.-H. Möller,
　　　　　T. Tran-Viet

189　Town Hall, Norderstedt
　　　competition

190　Johanneum sports hall, Lübeck
　　　competition, 1st prize

1981

191　Johanneum sports hall, Lübeck
　　　consultancy project

192　Kravag offices, Hamburg
　　　competition, 2nd prize

193　Kiel-Höm
　　　urban development expertise

194　Fontenay, Hamburg
　　　urban design expertise

195　Hotel, Lisbon
　　　consultancy project

196　Housing Abudja, Nigeria
　　　planning study

197　Nigerian Bank of Commerce
　　　and Industry, head offices, Abudja
　　　preliminary design

198　Municipal Library, Gütersloh
　　　competition

199　Refurbishment/conversion, Kiel Castl
　　　competition, 1st place

200　Law Court, Brunswick
　　　consultancy project, 1st place
　　　completed 1994
　　　D:　M. v. Gerkan with H.-E. Bock,
　　　　　M. Stanek, A. Buchholz-Berger
　　　P:　J. Zais
　　　C:　B. Kreykenbohm, H.-W. Warias,
　　　　　M. Skrabal, G. Wysocki

201　DAL office center, Mainz
　　　competition, 1st prize
　　　completed 1983/1992
　　　D:　M. v. Gerkan with J. Friedemann,
　　　　　G. Tjarks
　　　PL:　A. Sack
　　　C:　R. Henning, U. Rückel

202　Housing complex, Bad Schwartau
　　　competition, 1st prize

203　Apartment and commercial building,
　　　Marktarkaden, Bad Schwartau
　　　completed 1984
　　　D:　M. v. Gerkan with
　　　　　J. Friedemann, G. Tjarks

204　'Black Box' – Schauland sales hall,
　　　Hamburg
　　　completed 1982
　　　D:　V. Marg, K. Staratzke
　　　C:　O. Dorn

205　Low-Energy House, IBA Berlin
　　　competition, chosen for building
　　　completed 1984
　　　D:　M. v. Gerkan
　　　C:　M. Auder, P. Römer

206　Apartment and commercial building,
　　　Saalgasse, Frankfurt/Main
　　　completed 1988
　　　D:　M. v. Gerkan with J. Friedemann

207	Hotel Plaza, Hillmannplatz, Bremen competition, 1st prize completed 1985 D: M. v. Gerkan P: K. Brauer, K. Staratzke C: A. Buchholz-Berger, H. Nolden, H. Schmees, T. Tran-Viet	225	Inner city redevelopment, Niederes Tor, Villingen consultancy project	244	Quickborn block corner development competition, 1st prize
		226	Heimann Pickup wholesale store design	245	Film center, Esplanade, Berlin competition
208	Klinikum II, Nuremberg-South competition, 2nd prize	227	Parking concept, Celle consultancy project	246	Germanisches Nationalmuseum, Nuremberg competition, 2nd prize
209	Rama Tower Hotel, Bangkok design	228	Neumarkt redevelopment, Celle design	247	Ankara Kavaklidere Complex, Sheraton Hotel and shopping mall completed 1991 D: V. Marg, K. Brauer C: K. Lübbert, W. Haux, D. Heller, Y. Erkan, R. Preuss, D. Jungk, T. Bieling
210	Leisure park, Heiligenhafen consultancy project	229	Viktoria Insurance AG, Berlin consultancy project		
211	Institute of Chemistry, University of Brunswick competition, 2nd place	230	Savings Bank extension, Hamburg competition, commendation		
212	Connecting bridge between two office buildings, Hamburg case study	231	Daimler Benz AG offices, Stuttgart competition 2nd prize	248	Kunstmuseum der Stadt Bonn competition
213	Federal Postal Department, Bonn competition, 4th prize	232	Hillmannquartier, Bremen design	249	Intermediate school and sports hall, Schleswig-Holstein competition, 2nd prize
214	Komplex Rose, hospital for rheumatic diseases, Bad Meinberg competition, 1st prize completed 1988 D: M. v. Gerkan P: K. Brauer C: P. Römer, H.-R. Franke, H. Ritzki, B. Dziewonska, M. Stanek	**1983**		**1985**	
		233	Subway Station, Bielefeld completed 1991 D: M. v. Gerkan with H.-H. Möller	250	Various furniture designs D: M. v. Gerkan C: V. Sievers
		234	Hillmann-Garage, Bremen completed 1984 D: K. Staratzke C: P. Sembritzki, K. Lübbert	251	District Court, Flensburg competition, 1st prize completed 1989/1991 D: V. Marg P: K. Brauer C: B. Fleckenstein, C. Boesen, W. Haux, K. Lübbert, S. Schliebitz, S. Pieper
215	Central Post Office sorting halls, Munich competition, award	235	Refurbishment shopping mall, Hamburger Strasse, Hamburg consultancy project, completed 1988 D: V. Marg, K. Staratzke C: B. Fleckenstein, B. Gronemeyer, S. Peters, A. Lucks, H. Sylvester		
216	Sheraton Hotel, Berlin design			252	Post offices 1 and 3, Hamburg competition, 2nd prize
1982		236	Maria-Trost Hospital, Berlin competition, 4th prize	253	Wildlife Information Pavilion, Balje competition, 2nd prize
217	International Maritime Court, Hamburg case study	237	Health Insurance Corp., Hamburg competition, award	254	Reconstruction of Michaelsen Country House as a dolls' museum, Hamburg completed 1985 D (1923): K. Schneider D (1985): M. v. Gerkan, K. Lübbert, V. Rudolph
218	Stadtmarkt, Fulda case study	238	Gruner + Jahr publishing house, Hamburg competition, 1st prize		
219	Stollwerck-Passage, Cologne design consultancy				
220	Schlossparkhotel 'Orangerie', Fulda competition, 2nd prize	239	Housing Am Fischmarkt, Hamburg completed 1989 D: V. Marg M: M. Mews, W. Haux	255	Library, Göttingen competition, 5th prize
221	Apartment and office building Grindelallee 100, Hamburg completed 1987 D: M. v. Gerkan with K. Staratzke C: B. Fleckenstein, P. Sembritzki, H. Sylvester	240	Collegium Augustinum, Hamburg completed 1993 D: V. Marg P: K. Staratzke C: K. Lübbert, B. Fleckenstein, M. Dittmer, B. Gronemeyer, D. Winter, K. Heckel	256	Münster urban design competition, 3rd prize
				257	City Library, Münster competition, 1st prize group
222	Telecommunications office, Post Office 1, Regional Postal Directorate, Brunswick completed 1990 D: M. v. Gerkan PL: B. Albers C: K. Maass, A. Lucks, M. Ebeling, K. Lübbert, M. Mews, S. Pieper G. Tjarks, J. Friedemann			258	Bauforum, Hamburg
		1984		259	Interior structuring of Bertelsmann publishing house, Gütersloh consultancy project
		241	P.O. Parking Garage, Brunswick completed 1986 D: M. v. Gerkan PL: B. Albers C: K. Maass, A. Lucks	260	Pyongyang International Airport, North Korea projects A and D
223	National Library, Frankfurt/Main competition, award	242	Tourist information system, Hamburg case study	**1986**	
224	State and University Library, Göttingen consultancy project	243	Housing Fontenay, Hamburg competition, 1st prize	261	VIP State Pavilion, Pyongyang International Airport, North Korea design
				262	Labor Office, Oldenburg competition, 1st prize completed 1993 D: V. Marg

	P: K. Staratzke C: H. Huusmann, W. Haux, Y. Erkan, P. Zacharias, C. Kreusler, M. Ebeling, C. Papanikolaou		
263	Bäckerstrasse, Halstenbek, competition, a 3rd prize		
264	Hamburg Agency in Bonn completed 1986 D: V. Marg, K. Staratzke C: P. Römer		
265	Town Hall, Husum competition, 1st prize		
266	Kümmellstrasse, Hamburg competition, 2nd prize		
267	Concert hall for Lübeck		
268	Elbchaussee 139, Hamburg gmp · Bureau completed 1990 D: M. v. Gerkan C: J. Kienig, P. Sembritzki, S. v. Gerkan, V. Sievers		
269	Elbchaussee 139 – Le Canard completed 1990 D: M. v. Gerkan C: J. Kienig, V. Sievers		
270	Elbchaussee 139, von Gerkan residence, Hamburg completed 1992 D: M. v. Gerkan C: J. Kienig, V. Sievers, S. v. Gerkan		
271	Telecommunications Office 2, Regional Postal Directorate, Hanover completed 1994 D: M. v. Gerkan with K. Staratzke P: J. Zais C: G. Feldmeyer, J. Groth, S. Schütz, K. Pollex, T. Schreiber		
272	Sparkassenpassage, Linz, Austria design project		
273	Deutsche Lufthansa Training and Computer Center, Frankfurt/Main competition D: M. v. Gerkan C: J. Zais, H. Potthoff		
274	Museumsinsel, Hamburg competition, designs 1 and 2		
275	AMK trade fair halls, Berlin competition		
276	Bundeskunsthalle Bonn competition, 3rd prize		
277	Haus der Geschichte, Bonn, competition, 4th prize		
278	Seca factory and offices, Hamburg competition		
279	Free University, Witten/Herdecke urban design study		
280	Jumbo Shed, Deutsche Lufthansa maintenance hangar, Airport Hamburg project study		
281	Airport, Hamburg-Fuhlsbüttel competition, 1st prize completed 1993 D: M. v. Gerkan with K. Brauer		

344

C: A. Alkuru, T. Bieling, R. Dipper,
R. Franke, S. v. Gerkan, J. Hillmer,
K. Hoyer, F. Merkel, M. Mews,
T. Rinne, U. Schürmann,
C. Timm-Schwarz, P. Autzen,
K.-H. Follert, W. Gust, T. Hinz,
G. v. Stülpnagel

282 **National Theater, Tokyo**
international competition,
top 30 entries

283 **Technik III, Gesamthochschule Kassel**
competition, 2nd prize

284 **Police Headquarters, Berlin**
competition, 3rd prize

1987

285 **Municipal Library, Münster**
competition, 2nd phase, award

286 **Hillmannhaus, Bremen**
completed 1989
D: M. v. Gerkan and K. Staratzke
C: D. Papendick, B. Gronemeyer,
S. Dexling, A. Szablowski

287 **Civic Center, Wiesloch**
competition, 5th prize

288 **Research and lecture hall building,
Rudolf Virchow University Hospital,
Berlin**
competition, 2nd prize

289 **Biocenter, Frankfurt/Main**

290 **HEW Training Center, Hamburg**
completed 1990
D: V. Marg
C: H. Huusmann, P. Zacharias,
M. Ebeling, C. Kreusler

291 **BMW Customer Center, Munich**
design study

292 **'Kleines Haus' Brunswick**
competition

293 **Kleiner Schlossplatz, Stuttgart**
consultancy project

294 **Dockland development, Heiligenhafen**
competition, 2nd prize

295 **Birds' house**
design for TV guide 'Hörzu'
D: M. v. Gerkan

296 **Pfalztheater, Kaiserslautern**
competition, award

297 **Labor Office, Flensburg**
competition, 4th prize

298 **Telecommunications towers for
German Federal postal services**
competition, 2nd prize

299 **Löhrhof, Recklinghausen**
design

300 **Housing and commercial building
Moorbek Rondeel, Norderstedt,**
completed 1990
D: M. v. Gerkan with J. Zais,
U. Hassels
C: B. Kreykenbohm, A. Schwemer

1988

301 **Saar-Galerie, Saarbrücken**
consultancy project
completed 1991
D: V. Marg
C: H. Akyol, C. Hoffmann,
H. Nienhoff, J. Rind,
M. Bleckmann, B. Bergfeld,
R. Dorn, J. Hartmann-Pohl

302 **Speicherstadt
[dockland warehouse city], Hamburg,**
expertise

303 **Federal Ministry of Environmental
Protection and Reactor Safety, Bonn**
competition, 1st prize
final planning stopped

304 **Library of Technical University and
Fine Arts Academy, Berlin**
competition, 3 rd prize

305 **'Star Site', Birmingham,
International Business Exchange**
consultancy project

306 **EAM, Kassel**
competition, 1st prize
completed 1993
D: V. Marg, T. Bieling
C: Bieling + Bieling

307 **Zürich-Haus, Hamburg**
competition, 1st prize
completed 1993
D: V. Marg with N. Goetze
C: M. Mews, A. Lucks, S. Lohre,
T. Haupt

308 **Salamander building, Berlin**
competition, 1st place
completed 1992
D: V. Marg
C: J. Rind, M. Bleckmann,
P. Römer, S. Zittlau-Kroos

309 **Schering AG offices, Berlin**
competition, 4th prize

310 **German History Museum, Berlin**
competition

311 **Deutsches Luftfahrtmuseum,
Munich-Oberschleißheim**
competition

312 **Station square, Koblenz**
competition, 1st prize
at planning stage

313 **Multi-story parking garage,
Paderborn**
competition, 2nd prize

314 **Bank and commercial building
Brodschrangen/Bäckerstrasse,
Hamburg**
study
completed 1994
D: V. Marg, K. Staratzke with
S. Krause
C: D. Winter, C. Hegel,
J. Kalbrenner, K. Bonk,
S. Dexling, P. Sembritzki

315 **New Orangery, Herten**
competition, 2nd prize

316 **Störgang, Itzehoe**

317 **City Center, Hamburg 'Schenefeld'**
completed 1991
D: V. Marg, K. Staratzke
Pl.: B. Gronemeyer, A. Leuschner
C: K.-H. Behrendt, S. Bohl,
G. Feldmeyer, U. Gänsicke,
T. Grotzeck, K. Dorn

318 **Galeria, Duisburg**
completed 1994
D: M. v. Gerkan, K. Staratzke,
O. Dorn
C: M. Stanek, K. Krause, C. Zeis,
T. Grotzeck, J. Brandenburg,
H. Ladewig, E. Höhler

1989

319 **Jumbo Shed, Deutsche Lufthansa
Maintenance Hangar 7, Hamburg**
completed 1992
D: M. v. Gerkan, B. Brauer
P: M. Stanek, R. Nienhoff
PL: M. Engel, C. Schönherr,
D. Winter, W. Gust, G. Maaß

320 **Deutsche Lufthansa Workshops,
Hamburg**
completed 1992
D: M. v. Gerkan
Co: Pysall, Stahrenberg & Partner
Krämer
C: K. Staratzke

321 **Lufthansa Dock Store,
Airport, Hamburg-Fuhlsbüttel,** design

322 **Multi-story parking garage,
Hamburg-Fuhlsbüttel Airport**
completed 1990
D: M. v. Gerkan
PL: K. Brauer
C: K. Hoyer, U. Pörksen

323 **Leisure swimming pool, Wyk auf Föhr**
competition, 2nd prize

324 **Deutsche Bundesbank, Frankfurt/Main**
competition, 3rd prize

325 **International Maritime Court, Hamburg**
competition, 3rd prize

326 **Hertie Center Altona, Hamburg**
consultancy project

327 **Apartment and commercial building
Matzen, Buchholz**
completed 1991
D: M. Zimmermann with H. Peter
A: C. Perlick, K.-H. Schneider-Kropp,
C. Richarz, V. Sievers

328 **Savings Bank, Stuttgart**
competition

329 **Apartment and commercial building
Schaarmarkt, Hamburg**
completed 1994
D: V. Marg
C: H. Huusmann, D. Hillmer,
Y. Erkan

330 **Residential Park Falkenstein,
Hamburg**
competition, 1st prize

331 **Art Museum and town hall extension,
Wolfsburg**
competition, 5th prize

332 **Sternhäuser, Norderstedt**
completed 1992
D: M. v. Gerkan, J. Zais, U. Hassels
C: U. Wiblishauser, V. Warneke,
D. Engeler, T. Böhm

333 **documenta exhibition gallery, Kassel**
competition, 4th prize

334 **Glass roof, Museum of Hamburg
History, Hamburg**
completed 1989
D: V. Marg with J. Schlaich
C: K. Lübbert
Co: J. Schlaich

335 **International Forum, Tokyo**
competition

336 **Concert Hall, Dortmund**
competition, 3rd prize

337 **Extension 'Städtische Union', Celle**
competition, 2nd prize

338 **Carl Bertelsmann Foundation,
Gütersloh**
competition, 1st prize
completed 1991
D: V. Marg
PL: H. Schröder,
M. Zimmermann
C: H. Akyol, S. Jobsch

339 **Network Operating Station of
Preussen-Electra, Hanover**
consultancy project

340 **Railroad Station, Norderstedt-Mitte**
competition

341 **Television Museum, Mainz**
competition, 3rd prize

342 **VIP Lounge, Airport, Cologne-Wahn**
competition, 2nd prize

343 **Airport, Paderborn**
competition

344 **Bank and office center at Stuttgart
Main Station**
competition

345 **City Rail Station, Stuttgart Airport**
completed 1992
D: K. Staratzke
C: D. Perisic, B. Kiel

346 **Library,
Christian-Albrecht-University, Kiel**
competition, 4th prize

347 **Extension
Christian-Albrecht-University, Kiel**
competition, 5th prize

348 **Zürich-Haus, Frankfurt/Main**
competition, 5th prize

349 **Königsgalerie, Kassel**
competition, 2nd prize

350 **Harburger Hof, Hamburg**
at planning stage

1990

351 **Berlin-Tegel Airport,
parking garage P2**
completed 1993
D: M. v. Gerkan
P: R. Niedballa
C: P. Römer

345

352 Deutscher Ring, extension of office building, Hamburg
consultancy project

353 Deichtor + Ericusspitze, Hamburg
competition, 5th prize

354 Kehrwiederspitze – Sandtorhafen development, Hamburg
competition, 2nd prize

355 Office building, Mittelweg, Hamburg
competition

356 Deutsche Revision, Frankfurt/Main
competition, 1st prize
completed 1994
D: M. v. Gerkan
K: Staratzke
PL: A. Buchholz-Berger
M: M. Stanek,K. Krause,
G. Hagemeister,
J. Kaufhold, K. Maass,
B. Meyer, M. Engel,
M. Hoffmann, E. Grimmer

357 Neuer Wall 43, Hamburg,
Commercial Building
completed 1997
D: V. Marg
PL: T. Hupe
C: A. Alkuru, F. Hülsmeier,
D. Papendick,C. Berle,
D. Porsch,C. Timm-Schwarz,
A. Buchholz-Berger,U. Rösler

358 Music and Conference Hall,Lübeck
competition, 1st prize
completed 1994
D: M. v. Gerkan with C. Weinmann
PL: T. Rinne, W. Haux
C: V. Sievers, M. Klostermann,
B. Groß,C. Kreusler,
K.-H. Behrendt, P. Kropp

359 Hypo-Bank, Hamburg – 'Graskeller'
completed 1994
D: V. Marg
PL: W. Haux
C: J. Kalkbrenner, B. Staber,
R.Schmitz,K. Steinfatt

360 Landscaped apartment complex,
Kanzleistrasse, Hamburg-Nienstedten
competition, 1st prize

361 Technology Center, Münster
competition, 1st prize

362 Acropolis Museum, Athens
competition, 2nd prize group

363 Grosse Elbstrasse/
Carsten-Rehder-Strasse, Hamburg
design project

364 Münsterlandhalle, Münster
competition,3rd prize

365 Hotel, Palace au Lac, Lugano
design

366 'Zementfabrik', Bonn
competition,3rd prize

367 Neue Strasse, Ulm
competition, 2nd prize

368 Main building, Duisburg
consultancy project

369 Miro Data Systems, Brunswick
completed 1991
D: M. v. Gerkan
PL: U. Hassels, J. Zais
C: W. Gebhardt, U. Kittel,H. Timpe

370 Adult Educational Institute and
Municipal Library, Heilbronn
competition, 1st prize
planning stopped

371 Airport Center,
Airport, Hamburg-Fuhlsbüttel
case study

372 Aero-City, Stuttgart Airport
case study

373 Office center, Neuss-Hammfeld
case study

374 Subway station,Mönckebergstrasse,
Hamburg
completed 1991
D: V. Marg

375 Hillmann-Eck, Bremen
completed 1994
D: M. v. Gerkan, K. Staratzke, K. Voß
C: D. Porsch

376 Bertelsmann Press Building, Berlin
competition

1991

377 Atrium Friedrichstrasse, Berlin
consultancy project
completed 1997
D: V. Marg
PL: C. Hoffmann
C: B. Dieckmann,S.Djahanschah,
T. Naujack, E. Witzel,
H. v. g. Hassend, S.Küppers,
S. Ripp, S.Schindlbeck

378 Max Planck Institute for
Microbial Ecology, Bremen
competition

379 Café Andersen, Hamburger Strasse,
Hamburg
completed 1999
D: M. v. Gerkan
P: K. Staratzke
C: P. Sembritzki, O. Brück

380 Garden City on the Rhine, Speyer
consultancy and structural study

381 German-Japanese Center, Hamburg
competition, 1st prize
completed 1995
D: M. v. Gerkan
P: K. Staratzke
PL: R. Niehoff, K. Voß
C: A. Lucks, A. Perlick,
K. Rohrmann,K. Dorn,H. Eustrup

382 Kaufmännische Krankenkassen,
Hanover
competition, award

383 Office and commercial building,
Friedrichstrasse 108, Berlin
consultancy project
completed 1999
D: V. Marg
PL: C. Hoffmann,C. Hasskamp
C: F. Lensing, C. Dost

384 Railroad stations 'Rosenstein',
'Nordbahnhof', Stuttgart
competition, award

385 Stuttgart Airport, Tower

386 Development Elbe embankment,
Dresden
consultancy project

387 Business park in Brunswick
project study

388 BeWoGe Housing, Otto-Suhr-Allee,
Berlin
competition

389 Nordseepassage, Wilhelmshaven
completed 1997
D: M. v. Gerkan
P: K. Staratzke
PL: K. Falke, K. Heckel,M. Helmin,
M. Lucht, H.Münchhalfen,
D. Papendick,J. Rieger, K. Ritzke,
M. Sallowsky, R.Schmitz

390 European Trade Center, Brunswick
competition

391 Shopping mall, Langenhorn-Markt,
Hamburg
competition, 1st prize

392 Lenné-Passage, Frankfurt/Oder
competition, 1st place
under construction

393 Hansetor, Hamburg-Bahrenfeld
competition

394 Calenberger Esplanade, Hanover
competition, 1st place
completed 1999
D: M. v. Gerkan
P: N. Goetze
PL: K.Schroeder,
C: M. Klostermann,M. Haase,
T. Rinne, J. Steinwender,
C. v. Graevenitz,A. Perlick

395 Krefeld South II
urban design competition, 2nd prize

396 Altmarkt Dresden
competition, 1st prize

397 State Museum of the 20th Century,
Nuremberg
competition, award

398 Business Park,Münster docklands
consultancy project, 2nd prize

399 Salamander building, Peterstrasse/
Thomaskirchhof, Leipzig
consultancy project

400 Marina, Herne
competition, 2nd prize

401 Office towers, Frankfurt/Main
consultancy project

402 Office complex, Am Zeppelinstein,
Bad Homburg
competition, 1st prize

403 Süllbergterrassen,
Hamburg-Blankenese
competition

404 Neighborhood center,
Connewitzer Kreuz, Leipzig
consultancy project

405 Bei St. Annen/Holländischer Brook
development, Hamburg
at planning stage

406 HHLA office building, Burchardkai,
Hamburg
consultancy project

407 Harbour station,Süderelbe,
Hamburg
competition, 2nd prize

408 New Trade Fair, Leipzig
international competition, 1st place
completed 1996
D: V. Marg with H. Nienhoff
C: gmp team Leipzig

1992

409 Adult Educational Institute, Koblenz
planning stopped

410 Elbkaihaus, Hamburg
completed 1999
D: V. Marg
P: K. Staratzke
PL: S. Krause, A. Juppien
C: E. Hoffmeister, I. v. Hülst,
G. Venschott, D. Winter,
M. Ziemons

411 Housing Fontenay Allee, Hamburg
design

412 Operational Center, Stuttgart Airport
competition

413 Rütgers Werke AG, Frankfurt/Main
competition,3rd prize

414 EBL Office complex, Leipzig
preliminary design

415 World Trade Center, Berlin
competition

416 Spreebogen, Berlin
competition

417 Further Training Center,
Herne-Sodingen
competition, award

418 Cinema and shopping mall,
Harburg Carré, Hamburg-Harburg
planning stopped

419 Housing Friedrichshain, Berlin,
consultancy project
completed 1998
D: V. Marg
PL: S. Zittlau-Kroos, D. Heller
C: A. Foronda,I. Perkovic,
R.Scheurer, S. Winter

420 Forum Neukölln, Berlin
competition, 1st prize

421 Allee-Center, Leipzig-Grünau
competition, 1st prize
completed 1996
D: V. Marg
PL: S. Staratzke,W. Haux
B: B. Albers, K. Akay, H. Akyol,
K.-H. Behrendt, E. Grimmer,
M. Hoffmann,A. Lucks, K. Maass,
R. Preuß,H. Reusch,J. Rind,
K. Steinfatt, H. v. Szada,
L. Weinmann

2 Telekom operating building, Suhl
competition, 1st prize,
completed 1997
D: M. v. Gerkan with Bothe, Richter, Teherani 1992
D: M. v. Gerkan with J. Zais 1993
C: A. Reich, S. Dürr, U. Düsterhöft, H.-W. Warias, G. Wysocki, P. Weidmann, T. Böhm

3 Siemens-Nixdorf Service Center, Berliner Strasse, Munich
competition, 1st prize

4 Afrika-Haus, Hamburg
planning stopped

5 Mare Balticum Hotel, Bansin/Usedom
competition, 1st prize
planning stopped

6 Commercial building and library, Mollstrasse, Berlin
competition, 1st prize

27 Lecture Theatre Centre, University of Oldenburg
competition, 1st prize
completed 1997
D: M. v. Gerkan with K. Lenz
C: K.-H. Behrendt, B. Groß, S. Krause, B. Kottsieper, J. Reichert, D. Winter

28 District Court North, Hamburg
competition, 1st prize
planning stopped

29 Astoria Maritim Hotel, Leipzig
planning stopped

30 Apartment block, Schöne Aussicht, Hamburg
competition, 1st prize

31 Quartier Stadtsparkasse, development, Dresden
consultancy project, 2nd place

32 Domhof Municipal Theater, Osnabrück
competition

33 Emergency hospital, Dresden-Neustadt
competition

34 Sony, Potsdamer Platz, Berlin
competition, award

35 Daimler Benz AG, Potsdamer Platz, Berlin
competition

36 Olympia 2000, bicycle and swimming hall, Berlin
competition, 2nd prize

37 Ku'damm Eck Hotel, Berlin
under construction

38 Museum Türkenkaserne, Munich
competition

39 Cologne-Ehrenfeld urban design
competition, 2nd prize

40 Grundthalplatz, apartment block 10, Schwerin
competition

441 Noise control structures, Burgerfeld-Markt Schwaben
competition, 2nd prize

442 Prison, Hamburg-Billwerder
competition

443 Kongresshotel, Mannheim-Rosengarten
competition

444 Rehabilitation Clinic, Trassenheide, Usedom
consultancy project,
completed 1994
D: M. v. Gerkan with P. Römer
C: S. Zittlau-Kroos, R. Wolff, T. Herr, K. Baumgarten, H. Borgwardt, B. Galetto

445 Hapag-Lloyd offices, Rosenstrasse, Hamburg
consultancy project
completed 1996
D: V. Marg
P: K. Staratzke
PL: J. Jobsch
C: S. Winter, R. Preuss, K. Hoyer, W. Schmidt, H. Schöttler

446 Shopping mall, DSN-Terrain Stationsstraat/Honigmannstraat, Heerlen
consulting project

1993

447 Buildings for 'Premiere' TV channel, Studio Hamburg
completed 1994
D: K. Staratzke
C: K. Duncker, S. Krause, B. Gronemeyer, E. Werner

448 Hanseatic Trade Center, Kehrwiederspitze, stage IV, Hamburg
completed 1999
D: V. Marg
P: K. Staratzke
PL: H. Akyol,
C: A. Alkuru, K.-H. Behrendt, R. Dipper, K. Dorn, R. Giesecke, B. Gronemeyer, M. Lucht, T. Polakowski, C. Plog, U. Rösler, B. Scholl, E. Werner

449 Office building, Bredeney, Essen
competition, 1st prize
completed 1998
D: M. v. Gerkan
PL: W. Gebhardt
C: H. Gietmann, S. Kramer, R. Born, W. Höhl, C. Kreusler, T. Neeland, C. Papanikolaou, D. Papendick, C. Plog, H. Reusch, H. v. Szada

450 Main Park, Würzburg
project study

451 Labor Office Training Center of the Federal Republic of Germany, Schwerin
competition, 1st prize,
under construction

452 Police Headquarters, Kassel
competition, 3rd prize

453 Nürnberger Beteiligungs AG offices
competition, 2nd prize

454 Trabrennbahn, Farmsen, Hamburg
competition, award

455 Festival Hall, Recklinghausen
competition, 2nd prize

456 Hoffmannstrasse, Berlin-Treptow,
competition, award

457 Marienburg project, Nijmegen
competition

458 Dreissigacker-South urban design, Meiningen, Thuringia
competition, award

459 Secondary school, Crivitz
competition

460 Reichstag conversion for German Bundestag, Berlin
competition, 2nd prize group

461 City villa Dr. Braasch, Eberswalde
completed 1995
D: M. v. Gerkan
PL: P. Römer
C: R. Wolff, S. Derendinger

462 Lehrter Bahnhof – Central Station, Berlin
competition, 1st prize
under construction

463 Railroad bridge, Lehrter Bahnhof
at planning stage

464 Luisenstadt – Heinrich-Heine-Strasse, Berlin
competition, award

465 Redesign Hindenburgplatz, Münster
competition, 2nd prize

466 Museum Grothe, Bremerhaven
study project

467 Building for 'Der Spiegel', Hamburg
competition, 2nd prize

468 German-Japanese Center, Berlin
competition, 1st prize

469 Dortmunder Union Brewery
competition

470 Telephone offices 3 + 5, Telekom Berlin
completed 1998
D: M. v. Gerkan with J. Zais
PL: J. Zais
C: V. Warnecke, S. Schröder, S. Stodtko, G. Wysocki, P. Staak, S. Schütz, D. Schäffler, D. Rösinger, U. Köper, S. Schwappach, A. Schneider

471 Max-Planck-Gesellschaft, Marstallplatz, Munich
competition

472 Blankenese Railroad Station, Hamburg
urban design project

473 Railway Station, Berlin-Spandau
competition, 3rd prize
completed 1998
D: M. v. Gerkan
P: H. Nienhoff
PL: S. Zittlau-Kroos, B. Keul-Ricke, E. Menne
C: A. Schlüter, P. Bozic, A. Prebisz, M. Böthig, D. Berve, K. Baumgarten, M. Wiegelmann, G. Meyer, A. Dierkes, P. Wolf, M. Rothe, K. Struckmeyer,
P. Schuch, M. Stanek, W. Gebhardt, J. Kalkbrenner, B. Claasen, R. Lauer, B. Topper
Co: Schlaich, Bergermann und Partner

474 Railroad bridge across the Havel, Spandau Railway Station
completed 1998
D: M. v. Gerkan with J. Schlaich

475 Noise barrier, Train Station, Spandau
under construction

476 Spreeinsel, Berlin
competition

477 Neuriem-Mitte, Munich
urban design competition

478 Institute for Discrete Mathematics, Lennéstrasse, Bonn
competition, 4th prize

479 Erfurt-Ost, inner-urban extension
competition, 2nd prize

480 Footbridge, Holstenhafen, Lübeck
3 designs

481 Concert Hall, Kopenhagen
competition

482 Two Federal Department and Deutsche Telekom Mobilfunk buildings, Bonn-Beuel
consultancy project

483 Urban design "Stuttgart 21"
design project

484 Rosenstein Train Station, Stuttgart
study project

485 Trade Fair, Hanover, Hall 4
competition, 1st prize
completed 1995
D: V. Marg
P: K. Staratzke
C: T. Hinz, M. Ziemons, D. Vollrath, H. Ueda, B. Schöll, U. Rösler, K.-H. Behrendt, I. Pentland, R. Schröder, S. Winter, U. Heiwolt
Co: J. Schlaich

486 Telekom Training Academy, Klein-Machnow near Berlin
urban design project

487 Speicherblock X, Hamburg
at planning stage

488 Georgsplatz, Dresden
competition

489 Webergasse, Dresden
competition

490 Apartment and Commercial building, Deutrichshof, Leipzig
consultancy project, 1st place

491 Town Hall, Garbsen
competition

1994

492 EBL housing, Leipzig
planning stopped

493 Theater of the City of Gütersloh
competition, award

494 Housing Dovenfleet, Hamburg
consultancy project

495 Lecture Theatre Centre, extension of Chemnitz Technical University
competition, 1st place
completed 1998
D: M. v. Gerkan
PL: D. Heller, A. Lapp
C: A. Juppien, K. Maass, R. Schmitz, M. Watenphul

496 Extension University of Dresden
competition

497 Institute of Chemistry, University of Leipzig
competition, award

498 Stadtzentrum, Berlin-Schönefeld
competition, 1st prize

499 Kleist Theater, Frankfurt/Oder
competition, award

500 Bridge over the Horn, Kiel
consultancy project
completed 1998
D: V. Marg with J. Schlaich
PL: R. Schröder, H. Ueda, A.-K. Rose, D. Vollrath
Co: Schlaich, Bergmann und Partner, J. Knippers

501 Main Train Station, Leipzig
consultancy project

502 Forum Köpenick, Berlin
competition, 1st prize, completed 1997
D: V. Marg
PL: J. Rind
C: A. Harenberg, M. Kaesler, F. Lensing, G. Mones, M. Nowak, E. Sianidis, E. Witzel, R. Stuer, T. Behr

503 Federal Chancellery, Berlin
competition, 4th prize

504 Gerling Insurance Corp., residential and office villas, Leipzig
competition, 1st place
planning pending

505 Hypo-Bank, Frankfurt/Main
competition

506 Cardiff Bay Opera
competition

507 Tivoli Cinema, Berlin
planning stopped

508 Werdener Strasse, Düsseldorf
competition

509 Northern Wallhalbinsel development, Lübeck
competition

510 Business park on Robotron site, Sömmerda
consultancy project

511 mdr Middle-German Broadcasting House, Leipzig
competition, 4th prize

512 Holzhafen, Hamburg-Altona
competition

513 Apartment and Commercial block, Eppendorfer Landstrasse, Hamburg
competition, 3rd prize

514 Max-Planck-Institute, Potsdam-Golm
competition, 2nd prize

515 Building for cooperative Norddeutsche Metall-Berufsgenossenschaft, Hanover
competition, 1st prize, completed 1998
D: M. v. Gerkan
PL: W. Gebhardt
C: J. Kaufhold, A. Bartkowiak, C. Plog, B. Gronemeyer, M. Gorges, R. Giesecke, R. Blagovcanin, J. v. Hülst, L. Nachtigäller, C. Weitemeier

516 Train Station, Charlottenburg, Berlin
study

517 Standardized platform roofs in Saalfeld, Marktredwitz, Hof, Westerland, Wilhelmshaven, Mezzig, Kaiserslautern, Trudering, Bottrop, Eglharting, Wittenberg, Lübben, Mainz, Salzwedel, Ehlershausen, Bitterfeld, Bamberg, Lübbenau, Ulm, Freiburg, Frankfurt/M., Hamburg Railway Station, Hamburg Dammtor, Hagen, Treuchtlingen, Fürstenwalde, Angermünde, Velgast
and 40 at planning stage
completed 1998/1999
D: M. v. Gerkan, J. Hillmer
PL: V. Sievers
C: S. Bern, R. Dipper, B. Föllmer, M. Foudehi, M. Helmin, D. Hünerbein, J. v. Holtz, K. Nolting, A.-B. Springer, D. Tieu, P. Wedemann

518 Eastern Altstadtring, Dresden
competition

519 Berliner Platz, Heilbronn
consultancy project

520 Promotion Park, Bremen
competition

521 Gerlinghaus Am Löwentor, Stuttgart
commission
completed 1997
D: M. v. Gerkan with N. Goetze
PL: T. Grotzeck
C: C. Berle, K. Burmester, T. Haupt, E. Werner

522 Hauni AG Technical Center, Hamburg
planning pending

523 Helsinki Main Train Station
competition

524 Federal President's Office, Berlin
competition

525 Library of Zwickau Technical University
competition

526 Dorotheen Blocks, parliamentarians' offices, Berlin
under construction

527 Yokohama Harbour area
international competition

528 Redevelopment of AEG-Kanis site in Essen
competition, 2nd prize

529 Tiergarten Tunnel, Berlin
consultancy project
under construction

530 Development on Bahnhofstrasse, Erfurt
abandoned

531 Office building at the Stadtmünze, Erfurt
competition, 4th prize

532 University Library and urban design ideas competition, Erfurt University
competition, award

533 Airport, Zürich-Kloten
study

534 Main Train Station, Stuttgart 21
consultancy project

1995

535 Urban design Munich 21
design study

536 Main Railway Station, Munich 21
design study

537 Apartment and commercial building, Gerling Insurance Corp., Cologne
competition

538 C+L Deutsche Revision Wibera, Düsseldorf
competition

539 Extension Städtisches Museum im Simeonstift, Trier
competition

540 Dresdner Bank, Pariser Platz, Berlin
competition, 1st place
completed 1997
D: M. v. Gerkan
PL: V. Sievers
C: C. Abt, K. Dwertmann, P. Kropp, B. Queck, W. Schmidt

541 Redevelopment of former Domestic and Breeding Livestock Market, Lübeck
competition, award

542 High-Tech Center, Potsdam-Babelsberg
competition

543 Neumarkt, Celle
competition, 1st prize

544 Waterways and Shipping Board East, Magdeburg
competition

545 Extension Town Hall, Berlin-Treptow
competition

546 Elementary school, Munich
competition

547 Shopping mall, Berlin-Marzahn
study

548 Landwehrkanal Bridge, Berlin
under construction

549 Paper terminal, Pohl & Co, Kiel
competition

550 Main Train Station, Erfurt
competition, award

551 Museum 'Alte Kraftpost', Pirmasens
competition, 1st award

552 New Civic Center, Scharbeutz
competition, 3rd prize

553 'Cultural area' of Peat Bath Spa, Lobenstein
competition, 2nd prize

554 Trade Fair and Administration Center Bremen
competition, award

555 National Museum of Korea
competition

556 Police Headquarters, Frankfurt/Main
competition

557 Trade Fair, Hanover, Hall 13
competition, 3rd prize

558 New Forestry Teaching building, Dresden Technical University, Tharandt
competition

559 District Court, Brandenburg
competition

560 Hospital extension in Friedrichshain, Berlin
competition

561 University Hospital 2000, Jena
competition

562 Prison Kempten, Bavaria
competition

563 Potsdam Center – southern site (main station Potsdam-Stadt)
competition, 1st place

564 Town Center of Bahrenfeld, Hamburg
competition, 3rd prize

565 Apartment and business building with Markt-Galerie shopping mall, Leipzig
competition

566 Fachhochschule Rheinbach
competition

567 Multi-sports hall, Leipzig
competition, 4th prize

568 DSR Deutsche Seereederei, Christinenhafen, Rostock
planning abandoned

569 Stadthafen Rostock
competition

570 Fachhochschule Ingolstadt
competition,
urban design: 2nd prize
architecture: 5th prize

571 Parkstrasse roof cover, Wilhelmshaven
study

572 MP and Regional State Department buildings, Mainz
competition, 5th prize

573 Warnow-Passage, Rostock
consultancy project

574 Federal Labor Court, Erfurt
competition

575 Harare International Airport, Zimbabwe
study

576 Extension Klinikum Buch, Berlin
competition

577 New supermarkets complex, Göttingen
completed 1999
D: M. v. Gerkan, J. Zais
C: O. Schlüter, H.-W. Wanas, U. Düsterhöft, J. Ortmann, C. Thamm, G. Wysocki, H. Reimer, M. Geilenberg

578 Urban design Stuttgarter Platz, Berlin
consultancy project

579 Institute for Chemistry, Humboldt University, Berlin
competition

580 Espenhain, Leipzig
competition

581 Main Railway Station, Saarbrücken 21
consultancy project

582 Conference and Civic Center with hotel, Bochum
study

583 Gotha Main Station and station square
competition, 3rd prize

584 Dr. med. Manke consulting office, Uelzen
completed 1997
D: V. Marg, J. Zais
PL: J. Zais
C: D. Engeler, P. Staak, R. Duerre

585 Expo 2000 Railway Station, Hanover-Laatzen
competition, 5th prize

586 Museo del Prado, Madrid
competition

587 Living at Iseplatz, Hamburg
competition

588 Town Center of Schoneiche near Berlin
completed 1997
D: V. Marg
PL: M. Bleckmann
C: J. Hartmann-Pohl, F. Lensing, O. Drehsen, F. Jaspert, M. Kaesler

589 ICE 2.2 – train interior design
study and prototype

590 Railroad station of the future
at design stage

591 Holocaust Memorial, Berlin-Grunewald
design

1996

592 VIth Architecture Biennial, Venice Exhibition "Renaissance of Railway Stations. The City in the 20th Century"
D: M. v. Gerkan
C: D. Schäffler, S. Schütz, H. Tieben
conceptual system and design

593 Secondary school, Veits-Höchheim
competition

594 Expo 2000 – roofing over City-Railroad Stop, Hanover
completed 1998
D: M. v. Gerkan, J. Hillmer
C: K. Nolting

595 Prison, Gräfentonna
competition

596 HAB School of Architecture and Construction, Weimar
competition, award

597 Civic Center, Weimar
competition, 1st prize
completed 1998
D(1996): M. v. Gerkan with P. Kamps
D(1997): M. v. Gerkan with D. Schäffler and S. Schütz
PL: D. Schäffler, S. Schütz
C: K. Akay, M. Wiegelmann, H. Reimund, M. Böthig, A. Pfeifer, P. Bozic, J. Erdmann, P. Pfliederer

598 Residential Park Elbschloss, Hamburg
competition, 3rd prize

599 Urban design Bucharest 2000
competition, 1st prize

600 Hamburg Agency, Palais Luisenstrasse, Berlin
competition, 1st prize

601 Double bridge in Fürst-Pückler-Park, Bad Muskau
competition, award

602 German Foreign Office in Berlin
competition

603 mdr Middle-German Broadcasting Complex, Erfurt
competition, 4th prize

604 Broadcasting House, Thuringia
competition, award

605 New urban district, 'Layenhof/Münchwald', Mainz
competition

606 Commercial building in Riga/Latvia
design
at planning stage

607 Jurmala residence, Riga
completed 1998
D: M. v. Gerkan
PL: O. Dorn
C: J. v. Mansberg, J. Brauner

608 Nord LB Bank building, Friedrichswall, Hanover,
competition, 3rd prize

609 Urban design Frankfurt/Main 21
consultancy project

610 Main Railway Station, Frankfurt/Main 21
study project

611 Station 2000 – platform furnishings
under construction

612 Old railroad operating plant, Goslar
conceptual study 'locomotive depot'

613 Veterinary Faculty, Leipzig
competition, award

614 Metropolitan Express Train
completed 1999
D: M. v. Gerkan, J. Hillmer
PL: R. Dipper, B. Follmer,
C: S. Krause, F. Hulsmeier, M. Gorges, K. Kalb, B. Stehle, T. Neeland

615 Johannesburg Airport, SA
study

616 Housing Berlin-Lichterfelde
competition

617 Seating for regional trains, Deutsche Bahn AG
study

618 Ferry harbour, Mukran, Rügen
competition, 4th prize

619 Urban neighborhood center around Spandau Train Station, Berlin
consultancy project

620 Expo 2000 Plaza, Hanover
competition, 1st prize

621 Expo 2000 Plaza, Hanover
development study

622 Restaurant Vau, Jägerstrasse, Berlin
completed 1997
D: M. v. Gerkan with D. Schäffler, S. Schütz
C: G. Hoheisel

623 Main Railway Station, Bottrop
competition

624 Administration building, Am Wall, Göttingen
preliminary design

625 Extension of Hellerau Garden City
competition

626 DVG 2000 Administration, Hanover
competition, award

627 Main Railway Station, Darmstadt
competition

628 Spa Hotel, Hamm
competition, 2nd prize

629 Satellite apron west, Munich Airport
consultancy project

630 Schlossplatz, Berlin
study

631 Central University Library, Potsdam
competition, 3rd prize

632 Main Train Station, Lübeck
under construction

633 Tax Office, Schwarzenberg
competition

634 Transrapid Main Station, Hamburg
consultancy project

635 Haus Crange Training Hotel, Herne
IBA-competition, 2nd prize

636 Prison, Wulkow
competition

637 Fachhochschule Wismar
competition

638 Government buildings, Potsdam
competition

639 Private Residence, Alvano House
completed 1999
D: M. v. Gerkan, N. Goetze
PL: T. Haupt
C: G. Nunneman, N. Löffler

640 Housing Steinbeker Strasse, Hamburg
competition, 3rd prize

641 Landesfinanz-Rechenzentrum, Dresden
competition

642 Regional Parliament of Thuringia
competition

643 Station square, Kiel
preliminary design

644 Expo 2000 – 6 footbridges, Hanover
competition, 1st prize
completed 1999
D: V. Marg with J. Schlaich
PL: G. Gullotta
C: K. Reinhardt, B. Follmer, A.-K. Rose, T. Polakowski, M. Carlsen, S. Jöbsch, M. Ziemons

645 Housing and Commercial building, Altmarkt, Dresden
competition, 1st prize, under construction

646 Charlottenburg urban design, Berlin
competition

647 Central Bus Station, Oldenburger Stern
competition, 2nd round

648 Urban design, Ottakring Brewery, Vienna
competition, 1st place

649 Apartments, Am Stadtgarten, Böblingen
competition

650 Orthopedic Rehabilitation Hospital, Baden-Baden
competition

651 Urban design, Ostra-Allee, Dresden
competition, award

652 New building
for State Insurance of Swabia, Augsburg
competition

653 Gothaer Platz, Erfurt
competition, award

654 Fachhochschule Erfurt
competition, 3rd prize, award

655 Main Train Station, Kiel
under construction

656 Airport, Düsseldorf
competition, 4th prize

657 Technical University of Ilmenau
competition

658 Expo 2000 Plaza, Hanover
urban design masterplan

659	State Insurance head offices, Hamburg competition, award	680	Ryck Bridge, Greifswald consultancy project	
660	Airport, Zürich-Kloten competition, 1st place	681	Expo 2000, German Pavilion, Hanover competition	
661	Prison, Dresden competition	682	Apartment and commercial building, Erfurt competition	

659 State Insurance head offices, Hamburg
competition, award

660 Airport, Zürich-Kloten
competition, 1st place

661 Prison, Dresden
competition

662 Psychiatric hospital, Kiel
competition, 2nd prize

663 Trade Fair, Hanover, Hall 8/9
competition, 1st prize,
completed 1999
D: V. Marg with J. Schlaich and S. Jöbsch
PL: T. Hinz, M. Ziemons
C: A. Alkuru, M. Holtschmidt, K. Maass, S. Nixdorf, T. Schuster, A. Vollstedt

664 Conversion of Block D, dockland ware house city, Hamburg
consultancy project

665 Media-Center, Leipzig
competition

666 Deutrichs Hof, Leipzig
competition, 1st prize

1997

667 Thalia-Theater, Hamburg
restructuring/refurbishing
completed 1997
D: K. Staratzke
PL: D. Winter
C: M. Gorges

668 Bridge across the Wublitz
competition

669 Main Railway Station, Mainz
under construction

670 Railway Station, Limburg an der Lahn
competition

671 Builders' training yard, Berlin-Mahrzahn
competition, 2nd prize

672 Ford Research Center, Aachen
competition

673 Office and Commercial building, Am Kröpcke, Hanover
consultancy project
at planning stage

674 Shopping mall, Areal Kühne
study

675 Potsdam Center, facade design
competition, 1st prize

676 Foyer refurbishment, State Opera, Hamburg
competition, 3rd prize

677 Main Train Station Neighborhood redevelopment, Bielefeld
competition, 1st prize

678 Aiport, Kopenhagen
consultancy project

679 Dresdner Bank skyscraper, Frankfurt/Main
competition

680 Ryck Bridge, Greifswald
consultancy project

681 Expo 2000, German Pavilion, Hanover
competition

682 Apartment and commercial building, Erfurt
competition

683 Faculty building, University of Erlangen
competition

684 Multi-story parking garage, Trier
competition, award

685 Museum of Fine Arts, Leipzig
competition

686 New teaching building, Fachhochschule Weihenstephan
competition

687 Pediatric and Gynecology Unit, University Hospital, Dresden
competition

688 Connecting structures for Halls 3, 4, 5,6, 7 Trade Fair, Hanover
consultancy project
completed 1998
D: V. Marg
PL: D. Vollrath
C: T. Hinz, S. Hoffmann, M. Holtschmidt, B. Kottsieper, S. Nixdorf, U. Nibler

689 EVENT-Center, Essen
consultancy project

690 Living and Working on the Alsterfleet, Hamburg
competition

691 House Manke, Uelzen
preliminary design
under construction

692 Main Railway Station, Stuttgart 21
competition, award

693 Entertainment Center, Friedrich-Ebert-Damm, Hamburg
completed 1999
D: V. Marg, N. Goetze
PL: M. Bechtold, R. Schröder
C: C. Berle, T. Haupt, M. Carlsen
Co: Schild Architekten + Ingenieure

694 Synagogue, Dresden
competition, award

695 Renovation, Hapag Lloyd, Ballindamm, Hamburg
completed 1997
D: V. Marg, K. Staratzke
PL: D. Winter
C: K. Steinfatt, M. Gorges

696 Kassel Service Centre
competition

697 Potsdam Railway Station Quarter
competition, award

698 Old Airfield, Karlsruhe
competition

699 Town Hall, Saulgau
competition

700 Goree Memorial, Dakar, Senegal
competition

701 Federal Offices of Schleswig-Holstein and Lower Saxony in Berlin
competition, award

702 Horumersiel Clinic
competition

703 The Urban House, Berlin
competition

704 Bayerische Rückversicherung, Munich
competition

705 Trade Fair Rimini
competition, 1st prize
under construction

706 Chemnitz Industrial Museum
competition

707 IGA 2003 Rostock
competition, 1st prize
at planning stage

708 Grammar-School Waltersorfer Chaussee, Berlin
competition, 4th prize

709 Bad Steben Casino
competition, 1st prize
under construction

710 Trade Fair Düsseldorf
competition, 1st prize
under construction

711 Dresden Central Bus Station
competition, 2nd prize

712 Leipzig University - Humanities Faculty
competition

713 Constantini Museum Buenos Aires
competition

714 Transrapid, Interior Design
competition

715 Parish Centre, Johnsallee, Hamburg
competition

716 Bramsche Town Hall
competition, 2nd prize

717 Stuttgart Airport, Extension
under construction

718 Bremen, Teerhof
competition

719 Weserbahnhof II, Grothe Museum Bremen
competition, 1st prize

720 Christian Pavilion, Expo 2000, Hanover
competition, 1st prize
under construction

721 Linz Main Station
competition

722 Heckscher Klinik, Munich
competition

723 Design Hotel, Hamburg
competition

724 Elementary School Erkelenz
competition, 5th prize

725 Greifswald Bridge
consultancy project

726 Regensburg Burgweiting
competition, award

727 Stuttgart Airport, Terminal 3
competition, 1st prize
at planning stage

728 Basle Exhibition Grounds
competition

729 Elementary School Hanselmannstraße, Munich
competition

730 Police Headquarters, Technical Services, Duisburg
competition

731 St. Afra Grammar-School, Meißen
competition

732 Central Track Areas, Munich
competition

733 Institute of Physics, Berlin Adlershof
competition, 2nd prize

734 Deutsche Post AG Bonn
competition, award

735 Department of the Environment, Dessau
competition

736 Federal Offices of Brandenburg and Mecklenburg-Vorpommern in Berlin
competition, 1st prize
under construction

737 Berlin Brandenburg International Airport
competition, 1st prize
at planning stage

738 United Arab Emirates Residency, Berlin
consultancy project

739 United Arab Emirates Embassy, Berlin
consultancy project

740 Astron Hotel, Landsberger Allee, Berlin
study
completed 1999
D: V. Marg
PL: B. Lautz
C: D. Heller, K.-H. Behrendt, F. Möhler, J. Kaufhold, D. Rosinger

1998

741 Institute of Architects, Düsseldorf
competition

742 German School and Service Housing in Beijing
competition, 1st prize
under construction

743 Germany Industry Centre, Bucharest
planning stopped

744 Leipziger Platz Berlin
competition

745 International Congress Centre and Assembly Building Hanoi/Vietnam
competition

746 Technology Centre Bertrandt AG, Ehningen
competition, 3rd prize

747 Passau "New Centre"
consultancy project

| 8 | Federal Office of Sachsen-Anhalt, Berlin
competition, 1st prize | 774 | Trade Fair Shanghai
consultancy project | 801 | Mining Archives, Clausthal-Zellerfeld
under construction | 828 | Erfurt Authorities Centre
competition |
|---|---|---|---|---|---|---|---|
| 9 | Sparkasse Bremen
consultancy project | 775 | Egyptian Embassy in Berlin
competition | 802 | Stuttgart Schloßplatz
competition | 829 | Urban Villas Hamburg-Nienstedten |
| 0 | Munich Airport, Extension Terminal 2
competition, award | 776 | Jahreszeitenverlag, Hamburg
consultancy project | 803 | Quarter 115, Berlin-Mitte
consultancy project | 830 | Benetton, Hamburg
study |
| 1 | Trade Fair Amman, Jordan
consultancy project | 777 | Vocational Training School, Plattling
competition | 804 | Federal Department of the
Environment, Oppenheim
competition | 831 | Schirmdächer Expo 2000, Hanover
consultancy project |
| 2 | Caesar Foundation, Bonn
competition | 778 | Blankenese Train Station Square,
Hamburg
competition, 3rd prize | 805 | Development Concept Wieck/Eidena
study | 832 | Denzlingen Community Centre
competition |
| 3 | Music Theatre, Graz
competition | 779 | Residential Development Alstertal,
Hamburg
consultancy project | 806 | Regis Detention Centre
competition | 833 | Depesche Verlag, Geesthacht
study |
| 4 | Federal Office of Hessen in Berlin
competition | 780 | DB-Pavillon, Expo 2000, Hanover
study | 807 | Therapeutic Bath, Bad Kissingen
competition | 834 | Residential Building "Schiötz", Reinbek
at planning stage |
| 5 | C & A Site, Wilhelmshaven
consultancy project | 781 | Domplatte, Cologne
competition | 808 | Berlin Olympic Stadium, conversion
competition, 1st prize
at planning stage | 835 | Residential and Commercial Premises
Tacheles, Berlin
competition |
| 6 | Termina de Fusina, Venice
competition, 2nd prize | 782 | Concert Hall, Brügge
competition | | | 836 | Convention & Exhibition Center,
Nanning International, China
competition, 1st prize |
| 7 | Transrapid Station, Schwerin
competition | 783 | Villa Marta Extension, Riga | **1999** | | | |
| | | | | 809 | Renovation of Salzburg Main Station
consultancy project | 837 | Tempodrom, Berlin
competition, 1st prize
at planning stage |
| 8 | Bremen Rhodarium
competition | 784 | Apartment House for the firm
"Vincents", Riga | 810 | Renovation Innsbruck Main Station
consultancy project | | |
| 9 | Würzburg Sports Center
competition | 785 | Office Building for the firm "Vincents",
Riga | 811 | Schwäbisch Hall, urban design
competition, 3rd prize | 838 | Shopping Centre
Rothenburgsort, Hamburg
study |
| 0 | School, Vienna
competition | 786 | Congress Center, Rome
competition | 812 | Trade Fair Shenzhen
competition | 839 | Residential Buildings Nonnenstraße,
Leipzig
study |
| 1 | Donau Museum, Linz
competition | 787 | Future TV Station Beirut
consultancy project | 813 | Elbufer Dresden
consultancy project | 840 | Paderborn Airport
consultancy project |
| 2 | Entertainment Center Bielefeld
competition, award | 788 | Vienna Library
competition | 814 | German Embassy in Kiev
competition, 3rd prize | 841 | Trade Fair Hamburg
consultancy project |
| 3 | Bozen University
competition | 789 | Monument O'Connel-Street, Dublin
competition | 815 | Tromsoe Town Hall, Norway
competition, award | 842 | Ancona Airport
competition, 1st prize
at planning stage |
| 4 | Law Courts, Antwerpen
competition | 790 | Housing and Festival Quarter 54,
Wismar | 816 | Ku'damm 229, Berlin
consultancy project | | |
| 5 | Criminal Law Courts, Würzburg
competition | 791 | Bochum Bridges
competition | 817 | Schloßpassage Brandenburg
consultancy project | 843 | Dresden University
competition |
| 6 | College Refectory HTWK, Leipzig
competition | 792 | BASF Social Centre, Berlin | 818 | Vienna Airport
competition, 2nd prize | 844 | World of Sports "Adidas",
Herzogenaurach
competition |
| 7 | Administration for the Public Finance
Chamber, Berlin
competition | 793 | Library with Computer Centre
and Administration, Berlin Adlersdorf
competition | 819 | Bridges Expo 2001, Switzerland
competition | | |
| | | | | 820 | Telekom, Bonn
competition | 845 | Schloß Hopferau
competition, 1st prize |
| 8 | College Refectory, Regensburg
competition | 794 | Trade Fair Padua, Italy
competition | 821 | Thermal Wind Power Station,
Expo 2000, Hanover | 846 | Sports Hall Extension, Flensburg
at planning stage |
| 9 | Hotel Joachimstaler Platz, Berlin
consultancy project | 795 | Prager Straße Urban Design, Dresden
competition | 822 | Library 21, Stuttgart
competition | 847 | Maininsel, Schweinfurt
competition |
| 0 | Art Kite Museum, Detmold
competition, 1st prize
under construction | 796 | Museum, Göteburg
competition | 823 | Ludwigspassage Bamberg
study | 848 | Audi, Neckarsulm
study |
| 1 | Philips Convention Stand
competition, 1st prize
completed 1999
D: M. v. Gerkan with W. Haux and
M. Weiß
C: P. Radomski | 797 | Public Library Erbacher Hof,
Schweinfurt
competition | 824 | Biosphäre / Buga 2001, Potsdam
competition | 849 | Landshut Detention Centre
competition, award |
| | | 798 | New Trade Fair, Friedrichshafen
competition | 825 | Stuttgart Airport, Central Area
study | 850 | Production Hall STN Atlas, Hamburg
study |
| | | 799 | Neue Straße Ulm
competition | 826 | Diekirch Sports Center, Luxembourg
study | | |
| 2 | Airport Johannesburg
study | 800 | Münchner Tor
competition, 3rd prize | 827 | Museum, Schloß Homburg
competition, 3rd prize | | |
| 3 | Ulm Library
competition | | | | | | |

351